I0489851

Preface

The making of this book took nearly 30 years. This is the most comprehensive lottery book of its kind. I painstakingly ran through all the calculations. It works everywhere the lottery is played. Back test the book as far as possible. Lottery Icon gives you precision forecasting.

I would like to first give thanks to The Almighty God for using me as the medium to accomplish this book that a lot of people believed was impossible. I would like to thank my family and friends for being there with me through this difficult process of countless equations and predictions. I would like to thank you as well for giving me the reason to continue in my pursuit.

This book will put you outside the realms.

Thank you,

Eze Ugbor

Table of Contents

Chapter 1

- Solid Structural Foundation

Chapter 2

- Pick 3 Lottery Book

Chapter 3

- Pick 4 Lottery Book

Chapter 4

- Pick 5 Lottery Book

Chapter 5

- Lotto Winning Secrets Book

Chapter 6

- Becoming a Master (Practicum)

Lottery Icon

Chapter 1

Solid Structural Foundation

The name lottery board in this book implies the side that controls everything, including receiving all the proceeds from the bettors. They make the rules.

The only person excluded is the bettor, or a player, who happens to be the rest of us that bet all the money. We follow the rules made by the lottery board.

The rules set by the lottery board are based on the concept of luck. They say that the lottery is a game of luck. Luck in itself has two sides, good and bad. You may hope, like so many out there, to win the lottery at some point in your betting journey. The times you do not win, which tend to be many for a lot of bettors, do fall under the bad luck column.

A lot of bettors, however, do not even think of the bad luck days because they are overwhelmed with looking forward to the good luck days when they will enjoy their own winnings. The good days do elude so many, including the ones that win once in a while to cover some of the money they have lost in the course of looking for the good luck.

You will find a lot of those bettors among those who play the Pick 3 and Pick 4 lottery. If you are in doubt ask some of the bettors in the lottery line in your local store about the last time they won. You will find out that it has been a while for a good number of them.

Some of those players bet about twenty dollars or more for the midday, and that much for the evening game too. The quest for the good fortune removes them from factoring how much they have lost in the interim.

Since the lottery board sets the rules based on the concept of luck, I must remind the readers once more that the forces of luck do indeed fall under good and bad, and you are bound to take one of the two.

The contest is of course grabbing the good luck.

The good luck goes to the lottery board when you lose and comes to you when you win. The people making the rules are naturally going to write them in a way that benefits them. That is precisely why many lose.

There is, however, a small group that does win consistently regardless of the rules. The lottery board will make their money from those who lose.

Your goal then should be to belong to that small group that can win consistently.

There are people who still believe - to this day - that the lottery could not be won every single day. The small group who wins consistently will continue to do so regardless.

Would you like to be among the small group that enjoys consistent winnings?

I took the lottery games to court on behalf of the players that lost and are losing money from betting on the lottery numbers. The verdict came in your favor, to help you increase the opportunity of winning more often. The verdict is to write the final book you will ever need. Every single lotto game is covered in this book. The final verdict on your behalf is the ultimate book.

Lottery Icon.

There is no good fortune without effort.

You will be ready to read this entire book more than once.

This book is not written as a story. It is not going to be an easy read.

You will find information you have not come across before in this book.

I will equally use the methods I discussed in the number one lottery book,

Lottery Little Book

Chapter one will discuss in greater detail tables and trend formation. I do highly urge you to read chapter one before you proceed with the rest of the book.

The next chapter and the rest of the book will show you how to win every lottery game. At the end of each chapter, I will have some winning numbers for the readers who have limited time on their hand.

The convention on Pick 4 lottery is that the odds of winning are 1:10000.

This book will examine the limits of that convention.

I have always enjoyed working on Pick 4 with the idea of breaking the 10000 odd conventions.

I have done this over several years.

The other day I was going somewhere and noticed a vehicle with a unique tag number. I decided to put the tag number in the new method I developed to see what happens. It led me to the winning Pick 4 number 2424 in Maryland.

The next day, I visited somebody and decided to apply the name of the facility in the new method. It produced the next winning Pick 4 number.

I decided to go further by employing a completely different language in the equation.

Guess what happened?

It produced another solid winning Pick 4 number.

These good fortunes occurred one right after the other.

You are going to learn the above methods in this book.

This book is like learning a new language. You can achieve a lot more when you can speak the language. The lottery language that many are familiar with is the one where too many bettors lose.

The lottery language in this book is completely different from everything you know about the lottery.

Nearly every bettor has operated within the confines of what they learnt in their respective schools. That has not been adequate in the game of lottery betting.

The language in this book will teach you how to operate outside the realm. You will master how to reverse the game to your own benefit.

You can only do that if you make time and take effort to read and understand this book.

I challenge you to study this entire book and tell me that you cannot win Pick 4 and every other lottery game consistently.

You must master the tables and trends formation from chapter one.

The rules set by the lottery board are based on luck. If you follow the rules and lose, you must change the rules to win.

Be prepared to read this book.

Look forward to joining the small group who enjoy consistent winnings as long as you are ready to put forth the effort.

I have come to realize that some readers do indeed have some difficulty in figuring out how to develop and apply the tables. I will use most of this chapter to explain it. This is to make sure that everyone at the end of reading this book will be able to make profitable use of the tables.

There will be a lot to gain by reading this book in its entirety. This book is not going to be an easy read. It is not meant to be a novel.

You will, however, find it a lot easier to understand once you master how to create the tables.

Let us begin the rich process from this point on.

The simplest table, or what we call trend, will be the one below,

Step One

0	1	2
3	4	5
6	7	8
9	0	1

You will notice that the Pick 3 numbers in the rows are as follows,

012

345

678

901

They form a simple set of Pick 3 numbers in numeric order. I started in this simple order to make sure that the readers do follow along. It is very important that you understand this part before going to the other chapters.

The above groups are called a trend. As simple as they make look, they do follow a pattern.

This book will make great use of tables. You have to master it to be able to take advantage of the lottery game.

I am going to shift the above groups to one step further. I will do this from the numbers in the lead position (1st digits). The numbers are 0, 3, 6 and 9. I will continue in the same order from the middle set of numbers and finish with the numbers in the last digits.

Let me lay the entire group out before shifting the trend. The entire group in the order that I am going to shift them will be from,

012

345

678

901

To be changed into the ones below.

0, 3, 6, 9, 1, 4, 7, 0, 2, 5, 8 and 1.

I changed the group above by starting from the numbers in the first column beginning with 0, 3, 6 and continue with 9 and the groups in the middle and finally to the last digits.

I will equally try to explain the above with letters to help those who may have difficulty understanding it.

If you place the numbers in groups of three to create Pick 3 numbers beginning from the first digits the numbers will now look as

0, 3, 6, 9, 1, 4, 7, 0, 2, 5, 8 and 1.

When placed in Pick 3 format will appear as follows,

036, 914, 702 and 581.

You will now place the above newly formed Pick 3 numbers below each other as follows,

036

914

702

581

You will place this new group in the tables that will appear as follows,

Step Two

0	3	6
9	1	4
7	0	2
5	8	1

The groups when properly formed will go from:

012

345

678

901

To the new group:

036

914

702

581

I am going to at this point mix batch of letters and numbers to further clarify how I form the tables that create the trends.

The letters and numbers I will use for this example are,

A, B, C, 1, 2, 3, D, E, F, 4, 5, 6

I will use the above letters and numbers to create tables. Once I create the initial table, I will proceed to create additional three in similar order. What I am doing in essence is creating trends.

It is important that you understand this elementary process. I will go into more complex ones later in this book.

A, B, C, 1, 2, 3, D, E, F, 4, 5, 6

NL Tables

ABC	A1D	A4E	A35
123	4B2	31B	241
DEF	E5C	5FD	FCE
456	3F6	2C6	BD6

If you are creating just two tables in similar order with the following letters, AAA, BBB, CCC and DDD, you will form the original table in this order,

AAA

BBB

CCC

DDD

In order to create the next table or trend you will place the above letters in the order below,

ABC

DAB

CDA

BCD

And continue to create the tables or trends like the ones below,

AAA	ABC	ADC	ABD
BBB	DAB	BBA	BDB
CCC	CDA	DCC	CAC
DDD	BCD	BAD	ACD

Once more I accomplished this by taking the letters from the first column and form groups of three and continue into the letters in the middle column and finally into the third column.

If you look at the accompanying trends, they do not quite look like the groups that actually came from ABC.

This is one of the ways the lottery numbers move without being recognized by many. If you are trained to look at letters from the angle of ABC all your life, it will be difficult for many to recognize when the ABC trend produces the trend DAB and so forth.

The letters in the same rows fall under the same trend and will always exhibit similar behavior.

For instance, as you can see in the chart above, if BBB produced DDD, BDB will definitely produce ACD.

Ladies and Gentlemen please forgive me if the table formation up to this point seems repetitive or simplistic. This is to make sure that every reader understand it and follow along.

Believe me; we will get to the tough part. That part will be a lot easier if you understand the basic table formation.

For the readers who may not be clear at this point, I will replace the first three numbers from Step One with letters to show you how we created Step Two.

From this to This

012	A12
345	B45
678	C78
901	901

In the second part of the formation I replaced 036 with ABC. The next set of Pick 3 numbers will start from 9 after the ABC and continue in the same direction into the middle digits and through the rest of the group. This is how I formed Step Two.

I will use letters and other examples where necessary to explain table formations.

I will now put Step One and Two side by side.

Step One	Step Two
012	036
345	914
678	702

901   581

I will follow the same process and do Step 3 and 4 and place all four steps beside each other before proceeding.

012	036	097	050
345	914	531	493
678	702	086	827
901	581	421	161

As I stated so many times in this book, you will need to study the whole of this book to master all the tools necessary to enjoy consistent winnings.

 There is no great reward without effort.

I will enclose winning numbers at the end of each section of this book because I know that there are readers who will not read. I am doing that to justify their cost of purchasing this book. They are, however, being limited to the winning numbers that I have already worked out.

If you want to have the opportunity to win a whole lot more, be prepared to study the entire book. At the end of this book you will be able to work out trends that will have what we call 'zero inefficiency.' This is at the point you will be able to challenge the conventional odds of winning the lotto. You will be able to operate outside the realms.

Lottery Icon will employ a lot of examples from the DC metro area lottery market and bring in other states when necessary.

An example of what you will be able to achieve at the end of this book.

Actual Manual Lottery Trends against current Virginia lottery trends.

Manual Trend   Virginia Lottery Trend

Manual Trend	Virginia Lottery Trend
2553	2553 played 6/27/13
7124	7124 played 6/30/13
8409	4089 played 6/16/13
4674	7464 played 6/29/13
4512	1542 played 7/1/13

You can see from the above trends that the unique methods in Lottery Icon captured 100% of the Pick 4 numbers that won in less than one month.

In order to appreciate this level of efficacy, the 2553 and 7124 played straight. That would net about $10000 on $2 straight for the bettor that wagered on it.

You will be able to create this level of efficacy by the time you finish reading this book. Let the bystanders continue to argue that the Pick 4 and other lottery games cannot be won. Put your energy in studying and master the methods in this book.

You will need to study, not read.

You read novels.

You study and learn Lottery Icon because it is a language. People embrace you more when you can speak their language.

Lottery Icon is the actual language of winning the lottery. The reason: you will develop your skills to the level of trend calculation with zero inefficiency. At that point, the lottery winning numbers will embrace you consistently because you speak the same language.

You will be able to take actual lottery numbers that won and zero in on the next winning numbers. You will be able to create winning numbers and see when they are about to HIT.

You will be able to know how and when to create master trends that will be similar to the one above and capture the winning numbers.

Lottery Icon will leave no stone unturned.

If you are ready to study this book and create unlimited opportunities let the fun begin.

Chapter 2

Pick 3 Lottery Book

Now is the time to begin to expand upon the creation of the tables based on the safe assumption that you understood the basic table formations. If you are not conversant or familiar with the basic table formations please go back to them once more before you continue.

The table formation expansion is necessary because the lottery plays a variety of trends. You need more than the basic trend to capture the winning numbers consistently.

The numbers I will be using to create the tables will be numbers 0(zero) through 9 (nine).

All lottery numbers will fall between 0 and 9.

I will introduce the three methods that are in the number one best-selling lottery book,

Lottery Little Book.

The three methods are necessary in the development of master trends with zero inefficiency. I will explain later in this book.

I will at this point create an expanded table. I will give you the finished table and show you the steps to create the table.

 Lottery Icon is the practicum of everything you will ever need in order to win the lottery.

When finished, the first expanded table will look like the one below.

Finished Table.

4	3	8	5	4	8	2	0	2	5
4	7	2	3	4	5	9	1	6	1
6	5	2	1	3	8	6	7	0	1
6	1	6	4	0	9	8	7	2	9
0	7	5	7	3	9	0	3	8	9

I will now begin to show you how to create the above table. The numbers inside the table are 0 through 9.

Expanded Table Formation 1.

Step One

				4					
			3		5				
		2				6			
	1						7		9
0								8	

You will notice in the above table how I placed the numbers 0 through 9. The process will continue from zero (0) once you get to number 9 every time until you fill the entire grid.

Step Two

				4		2			
			3		5		1		
		2				6		0x	
	1						7		9
0								8	

You will notice that I placed an 'x' beside the 0 that is the continuation of the process. You will be able to only get numbers 0, 1 and 2 on the top right hand side in this particular process.

The next occurrence of the number 3 will continue in the same direction. You will jump over the number 5 to continue. Please remember that you will do the same each time you get to a cell in the table where a number is already in place.

The next table will look like the one below once you continue with number 3 and so forth.

Step Three

				4		2			
			3		5		1		
		2		3x		6		0x	
	1		4				7		9
0		5						8	

I placed an 'x' beside number 3 to show you where the next number continued from. In Step Three you now have the second round of numbers up to 5. You will need to continue with number 6. In order to do this, you will jump over number 1 and continue with 6. I will place an 'x' beside number 6 to show the readers the continuation of the number placement in the grid in Step Four.

Step Four

		8		4		2			
	7		3		5		1		
6x		2		3x		6		0x	
	1		4				7		9
0		5						8	

I have now placed the numbers 6, 7 and 8. In order to continue, the next number will be 9, after which the process will begin from 0. The number 9 could only be placed after number 3x to be able to continue.

That will now bring us to Step Five, at which point the table will now look like the one below.

Step Five

		8		4		2			
	7		3		5		1		
6x		2		3x		6		0x	
	1		4		9		7		9
0		5				0		8	

I have now placed the numbers 9 and 0. I will hasten the process by placing the remaining numbers with the letter v beside them. In the above table there are only four spaces left to finish one half of this process.

Since I have just finished placing 9 and 0, I will continue with the remaining four, which will be the numbers, 1v, 2v, 3v and 4v.

I will remove the letters at the end of the process. They are meant to help you during the formation of the tables.

The Step Six table will look like the one below,

Step Six

4v		8		4		2		2v	
	7		3		5		1		1v
6x		2		3x		6		0x	
	1		4		9		7		9
0		5		3v		0		8	

You have now completed one half of the table.

The second half is very simple. If you look at the above table, you will notice that you can fill the rest of the grid without jumping any letter. The last number you placed was 4v. You will now continue with number 5 on the top right corner of the table. That top right corner is to the right of 2v and above 1v.

I will just place numbers on one end from the top right corner before continuing to finish the table creation. Do not forget that I am starting with number 5 since the last number I placed was 4 (v).

Step Seven

4v		8		4		2		2v	5
	7		3		5		1	6	1v
6x		2		3x		6	7	0x	
	1		4		9	8	7		9
0		5		3v	9	0		8	

You can see that I placed numbers 5 through 9. You will continue in a similar order until the table is filled. At this point, I will continue to fill in the rest of the grid. The next number will be 0. If you place it correctly, the 0 will be above 3v and below 3x. I will continue from Step 8. It is very important that you understand this process before you proceed.

Step Eight

4v	3	8		4		2		2v	5
4	7	2	3		5		1	6	1v
6x	5	2	1	3x		6	7	0x	
	1	6	4	0	9	8	7		9
0		5	7	3v	9	0		8	

You have now filled the grid up to number 7. In order to continue, you will have to jump over the number 0. Remember that the 0 is the one above 3v and below 3x that started this round. I will now continue in Step Nine with number 8 since I just finished placing number 7.

Step Nine

4v	3	8		4		2	0	2v	5
4	7	2	3		5	9	1	6	1v
6x	5	2	1	3x	8	6	7	0x	
	1	6	4	0	9	8	7		9
0		5	7	3v	9	0		8	

I will assume that you pretty much understand how to do this at this point. In order to continue, I will jump over the number 6 on the right side; below 2v and above 0x. I will fill out the rest of the grid in Step Ten beginning with the number 1 since the last number I placed was 0.

If you are not quite sure go back to prior table and study it carefully. Each successive table is a continuation of the last one.

Step Ten

4v	3	8	5	4	8	2	0	2v	5
4	7	2	3	4	5	9	1	6	1v
6x	5	2	1	3x	8	6	7	0x	1
6	1	6	4	0	9	8	7	2	9
0	7	5	7	3v	9	0	3	8	9

Once you remove the letters, the finished expanded table will look like the one below.

Finished Table

A	B	C	D	E	F	G	H	I	J
4	3	8	5	4	8	2	0	2	5
4	7	2	3	4	5	9	1	6	1
6	5	2	1	3	8	6	7	0	1
6	1	6	4	0	9	8	7	2	9
0	7	5	7	3	9	0	3	8	9

Now that you have finished building your first table, let us take a look at one small section of the table to make sure that the trends do indeed work. If you construct this table incorrectly, it could cost you big money.

It is important that you get it right.

Let us look at the Pick 3 from columns A, B and C against actual Maryland lottery Pick 3 results. If it works in one state, it will work in every other market.

Maryland lottery played the first Pick 3 numbers from the table 438 as 483 on 5/13/2013 as well as 616 that played as 661 on the same day.

It is indeed interesting to know that two Pick 3 numbers you just finished forming played the same day under the same columns.

The two Pick 3 numbers that won played in the exact same way, by the way.

The numbers we created are 438 and 616. Maryland lottery played 483 and 661. If you shift the middle digits to the third digits, they will form exactly the same numbers Maryland played.

Some people might be tempted to wonder if this is some kind of coincidence.

You should be able to satisfy their curiosity by pointing out to them that Maryland played 652 on 5/31/2013 and 472 on 6/1/2013. In this case both Pick 3 winning numbers played in the exact format that our table produced within a period of two days.

It is important to note that the groups in the above table will not play in the exact same way in every lottery market. The demonstration above is to show you that you indeed created your first lottery number trend that will play anytime your state lottery starts playing the trends. In the true sense of the lottery game, there is no other way the lottery numbers play, beyond trend.

The above two winning Pick 3 numbers, 652 and 472 will pay the bettor about $500 for each $1 wagered. That should be considered a very good return on a $1 bet.

The last group of winning Pick 3 numbers is 075, which played as 705 and 057 on 7/18/13 and 7/24/13 respectively.

Lottery Icon

It is not too late to win with the above groups. The winning numbers from this table you created are coming to the lottery market near you.

Do not forget that the remainder of the groups in the table exhibit similar behavior, which means that you stand to reap the reward if you follow the trends carefully.

The first four winning Pick 3 numbers in the group played in a period of about two weeks. That feat should make every patient lotto player happy.

The real vacuum that needs to be filled is the question of the days in between the two weeks it took for 80% of the group to play.

There are winning Pick 3 numbers every day.

Your job as a master is to catch a chunk of those winning numbers.

Are you still in doubt that lottery numbers can be won every day?

Be prepared to study this book hard and put into practice everything that you will learn from it.

You will have to put in some work, like with every other worthy endeavor in life.

Be prepared to read this book several times if necessary.

Your motivation ought to be on consistent winning lottery numbers.

Let me create one additional table for the readers who may not be very clear on the explanations we have done so far.

I will use two tables for this exercise by placing one half of the entire sentence in one and complete the rest of the table with the other half.

I will begin by placing the letters from the lower left side of the table and continue through the rest of the process in the same manner I have been explaining.

I will use an actual sentence below to show you how to create the tables.

I will study this book very well and make money with it, period.

L		V		L		I		W	
	K		L		S		H		Y
O		I		S		T		T	
	W		B		E		U		Y
I		O		E		R		D	

In the above table I placed one half of the above sentence.

I will continue the rest by starting from the top right side of the table below. Please bear in mind that you can start from any part of the table in order to complete it. I chose to begin the second half from the top right side.

I will place an asterisk * symbol to show you where I will start the second part.

	M		E		O		T		L*
O		E		P		I		A	
	N		K		W		N		H
R		E		A		D		I	
	I		Y		M		T		D

I have now finished placing the second half of the sentence,

<u>I will study this book very well and make money with it period.</u>

The completed table that will encompass the entire sentence will look like the one below,

L	M	V	E	L	O	I	T	W	L
O	K	E	L	P	S	I	H	A	Y
O	N	I	K	S	W	T	N	T	H
R	W	E	B	A	E	D	U	I	Y
I	I	O	Y	E	M	R	T	D	D

This is just one more step towards understanding how to create the tables.

If you are ready to start digging deeper and mastering how to make more money through lottery games, continue on to the next chapter.

Pick 3 Lottery

Pick 3 lottery is one game that does not fall under the laws of probability and yet it is still elusive to many.

Lottery Icon is not just going to mitigate, but remedy the situation.

You will develop the faculty to generate consistent winning Pick 3 numbers by the time you finish reading Book 1.

Before I continue on the Pick 3, let me touch briefly on one lucrative area of lottery games: Pick 2 lottery. The Pick 2 lottery is the least known and promoted lottery game, yet it yields good money for the few who understand it.

The actual lottery jargon is parlay. You properly play it as a front and back pair. It might be called other names elsewhere, but the most important thing at this point is to understand it, employ it in your betting games, and make money with it.

If you are betting parlay on the Pick 3 number '223,' the way to bet the parlay, or Pick 2, will be by playing,

X22

22X

X23

23X

X32

32X

In the above scenario, the parlay gave you six different results. The X represents 0 through 9. It does not matter what number plays in place of X; you will win.

A $1 parlay bet on the 223 will cost you a total of $6 and you stand to win about $50. You can bet on the 223 as well as the parlay and equally win on both.

If, on the other hand, you wagered on just 223 and your local lottery market played 213, you will end up losing unless you played the parlay.

The parlay is meant to often return your cost of betting, and more, for the next time around. Any lottery master should be looking for ways to infrequently bet with their own money.

The house does not lose from many. You should be among the few that take advantage of lottery trends and enjoy consistent winnings.

If, in the above example, you are betting parlay on 213 instead of 223, you will have a total of twelve front and back pair numbers to bet on.

You can equally bet 50 cents and capture $25 when you win. That will be more than enough to cover your cost of betting and to stay in the game until the big winners.

Using the DC Pick 3 lottery, allow me to demonstrate some opportunities with the trend table you have just created (Finished Table).

In the table, let us look at columns C, D and E.

The Pick 3 numbers in those columns are,

C	D	E
8	5	4
2	3	4
2	1	3
6	4	0
5	7	3

Let me put the above numbers side by side with actual DC 3 lottery numbers.

Table   DC 3

854   456 played on 1/7/13. You need only two common numbers among the two
(45) to win.

234   236 two common numbers 23 to win the parlay.

213   236 on 1/8/13 and 321 on 1/10/13

640   456 on 1/7/13 the common numbers among both are 6 and 4.

573   576 on 1/8/13 the common digits are 57 to win.

The same parlay trend continued in the DC lottery from 1/9/13 through 1/12/13.

I will now put the DC winning numbers beside those from our table from columns D, E and F

Finished Table and DC 3 results that are common and would have produced winnings.

Table	DC 3
548	984 and 484
345	456 would have won two days prior.
138 two	321, 308 and 893 would produce winnings on the parlay bet. You can see the common digits among them.
409	984 has two common digits among them.
739	739 would have paid off handsomely on the parlay, box and straight.

There are some instances where the two common numbers will not be together like the 409 and 984 above. This instance will make it difficult for those who did not study the Lottery Little Book and or Lottery Icon to win those.

The reason for that is because despite all your good intentions up to this point, the above table has gaps.

The gaps are there precisely in order for you to lose.

If you have not paid close attention, this is the time for you to begin to do so.

At this point you may want to take a break, digest what you studied so far - perhaps even drink a cup of tea.

Now, let us touch on those gaps before embarking on the heavy trends with zero inefficiency. You are about to start the journey that will place you ahead of the crowd.

At this point you have taken a well deserved rest and are back to the book, ready to put in some real work.

Finished Table

A	B	C	D	E	F	G	H	I	J
4	3	8	5	4	8	2	0	2	5
4	7	2	3	4	5	9	1	6	1
6	5	2	1	3	8	6	7	0	1
6	1	6	4	0	9	8	7	2	9
0	7	5	7	3	9	0	3	8	9

Let us take a quick look at the above Finished Table through the lens of the Virginia Pick 3 Lottery.

On 6/16/2012, Virginia played 897, which took me to the fourth row of column F, G and H. You will find the winning number 987. From this starting point, I will be watching for the next winning number to figure out the direction of the trend so that I can participate in getting some winning Pick 3 numbers.

Your goal should always be to see winning numbers with at least two common digits in order to confirm the trend before putting your money in the pot.

In the next result Virginia played 609. In this trend you have 903 below the 987 from column F, G and H.

You will find two common digits among the 609 that they played and the 903 that our Finished Table produced.

The next winning Pick 3 numbers that Virginia played were 584 on 6/17/2012. In this case you would have lost the bet.

The Pick 3 number that won is in the Finished Table, but did not follow the trend, at least for the moment, which you are betting on.

Based on the 987 and 903, you would have bet on 820 and 867 in the same column.

The new dilemma is that they played 584 from columns C, D and E.

You are now saddled with the choice of continuing to play the winning numbers you have committed to from columns F, G and H or pursue the new trend forming with 584 from columns C, D and E.

I am not saying by any means that the winning Pick 3 numbers from column F, G and H are not going to play. They are going to play based on the fact that 987 and 903 share at least two common digits with 897 and 609 that Virginia played.

The next day 6/18/2012, Virginia Pick 3 played 338 and 827.

The patient bettor who continued playing the numbers from columns F, G and H would have won the parlay bet on 827 with the winning number 820. This effort took an additional one day but turned out to be profitable, even if the profit is very minimal.

The number one rule in betting is preservation of capital. You are not going to bet for very long in this game if you do not take rule number one seriously.

The same winning number 827 that paid a few dollars into your pocket generated its' own trend from column G, H and I. It actually played as an extension of 897.

The 827 and 338 played on 6/18/2012. Our Finished Table, on the other hand, has the winning numbers 872 and 038 right below it on columns G, H and I.

You will notice that there are at least two common digits among them. You have 872 from our table Pick 3 winning numbers, 827 and 038, against Virginia Pick 3 winning number 338.

You likely would have missed the two Pick 3 numbers because there was no trend prior to the two playing. The two winning numbers, however, established a trend. This means that you could pursue the third winning number from the new trend. The two new potential winning numbers based on this trend would be 670 and 202.

The 670 and 202 will be the winning Pick 3 numbers right above and below the established trend from columns G, H and I.

You would have won the parlay in the next game because Virginia lottery played 120.

The other winning Pick 3 number, 854, from columns C, D and E, equally produced some winning numbers. If you had embarked on playing the rest of the group in that column, you would have spent a little bit more.

The reason being that the winning numbers did not play right after each other. The winning numbers would have been 213 matched against 120, and 640 against 609.

The 385 played exactly like the 827. You will notice that it is an extension of 854, just the same way 872 is an extension of 987.

You would win on the 853 if you followed the trend that created it. In that trend you will find from column B, C and D winning Pick 3 number, 723, matched against 827, and 521 against 120, which would leave you with the choice of betting on 385 and 757.

In this case the trend started from the middle and continued outwards. You must remember that the Finished Table is like a circle and the winning numbers will consistently rotate within the circle.

I highly encourage you to look at the trend that produced the 385 once more in order to note that the trend could start from any part of the table and must be recognized to take advantage of the opportunity.

In every one of these instances, they did not produce winning straight numbers. You could have made some money but not quite as you would like.

You must not forget to master the trend observation from all angles of the table.

I will give one instance from columns E, D and C.

If you look at each one of those columns beginning from column E, you will notice the Pick 3 numbers 430.

The next winning number from column D, in the same position, will be 314 and the next column beside that will be column C, with the winning number 226 in the same position.

You might end up not seeing the trend unless you develop the keen interest.

Let us look at the winning numbers that matched at least two common digits with those from Virginia.

 The winning Pick 3 number 430 matched Virginia's Pick 3 winning number 934 that played on 6/25/12.

The winning Pick 3 number 314 matched Virginia's Pick 3 winning number 614 that played on 6/26/12.

The winning Pick 3 number 226 matched Virginia's Pick 3 winning number 652 that played on 6/27/12.

The above winning Pick 3 trend played in midday results.

While you are observing and looking at the midday trend, you should keep a good eye on the evening results too.

Let me show you trends for the evening in the exact timeframe that will produce winnings for you in both midday and evening.

Take a good look at columns H, I and J on the same dates that produced the above winning numbers.

The lottery people are not in the business of making it easy for you, which is why I am putting forth the effort to show you how to identify trends and make consistent winnings.

In our finished expanded table, we have the Pick 3 numbers 199 in column J, 028 in column I, and 773 in column H. You will find all three of the winning Pick 3 numbers beside each other in exact positions.

Let us put the above winning numbers against Virginia lottery results covering the same dates to produce at least two common digits.

The winning Pick 3 number 028 matches the Virginia Pick 3 winning number 829 that played on 6/25/12.

The winning Pick 3 number 199 matches the Virginia Pick 3 winning number 921 that played on 6/26/12.

The winning Pick 3 number 773 matches the Virginia Pick 3 winning number 773 that played on 6/27/12.

In the above instance, they started with 028 for 829 in column I, moved rightwards to 199 of column J for winning Virginia Pick 3 numbers 921. A lot of readers will go in the same direction and pick the next potential winning Pick 3 numbers 660.

I always advocate choosing the Pick 3 numbers on both sides of the trend.

They played column I, J and completed it with column H that produced the straight winning Pick 3 number 773.

As you can see, the Virginia Pick 3 results produced at least two common digits.

You would have been able to win on the 652 that played in the midday drawing as well as the 773 in the evening.

Let me bring down the finished table for those readers who may find it difficult to scroll up and check the table against the analysis we have done so far.

I will as well check it against one more state result to demonstrate the gaps as well as inject any necessary questions that would arise at this point.

Every reader who thinks that the lottery could not be won every single day, please pay close attention from this point on.

Finished Table

A	B	C	D	E	F	G	H	I	J
4	3	8	5	4	8	2	0	2	5
4	7	2	3	4	5	9	1	6	1
6	5	2	1	3	8	6	7	0	1
6	1	6	4	0	9	8	7	2	9
0	7	5	7	3	9	0	3	8	9

Let us take a quick look on the above table through District of Columbia winning Pick 3 numbers. I will like to use the winning numbers from the same dates.

Through similar initial dates (6/16/12 through 6/19/12),

DC 3 played winning Pick 3 numbers,

855

260

403

101

245

521

811

937

There were two friends, one each from DC and VA, the friend that was betting on the DC 3 would start to smile with his Virginia betting friend because both parties started with identifying winning trends.

They embarked on a similar journey from different states that gave the two friends winning Pick 3 numbers on 6/16/12.

They might have gone to a good restaurant near the zone line to celebrate in anticipation of winning big, since they identified their respective trends before placing the bets.

You will find DC 3 winning trend in our table in columns E, F and G.

DC 3 played the columns as F, G and E that produced from our table 858 for 855, 296 for 260 and 403 for 443.

You will notice two common digits between the two parties that produced the trend, thereby giving the DC friend a winning parlay number in 443.

The readers who might have forgotten by now should note that he would place the parlay bets for the 443 as follows,

X44, 44X, X43, 43X, X34 and 34X.

There is, however, one problem with the numbers that the DC friend wagered on. The trend he is betting on has some gaps.

You will notice that DC 3 played the winning Pick 3 number as 403. The newly introduced number 0 (zero) produced the gap. Despite the fact that the trend has two common digits, he lost because of the gap. He promptly cut his broad smile short.

He is now starting in a hole and needs to recoup the money he spent in the restaurant.

You will find the winning DC 3 numbers that played in column E as an extension of the 443 he placed his bet on. Right beside the 430 of the Finished Table, you will find the winning number 589 that satisfied the trend for 855 that DC 3 started with.

The next winning DC 3 number after 855 that played on 6/16/12 is of course 260. The winning number 260 is placed all the way in column I of the Finished Table.

This scenario violated the trend rule, because the DC friend had no way of knowing that 260 is going to play after 855 for the simple fact that the two winning numbers are not placed side by side. This is indeed a dilemma for our DC friend.

But he is not giving up. He has now noticed that 101 played after 403 on 6/17/12. There is a trend that he might place his hands on. You will find in column I and J the winning numbers 260 and 511 as well as 602 and 119.

The DC friend placed his bet based on the newfound trend and lost again because DC 3 played 245 and 521 on 6/18/12.

This must be excruciating because his friend is laughing all the way to the bank while he is losing money. He is no longer eager to stop by the restaurant. He is angry to the point of not talking to his friend.

How can he be listening to this guy screaming about steady winnings and not having any luck? What could be the problem?

I don't know about you, but a lot of people will not find it funny either!

You will find the winning DC Pick 3 number 521 in the fourth row of columns B, C and D.

This number played straight just like the 260 that he also missed.

There was no way he could have gotten the 521.

DC 3 is producing consistent losing numbers because of the gap that was created from 403 against 443.

If he is looking at using DC 3 to make consistent part time money, he will be highly discouraged.

There are winning DC 3 numbers for him in the Finished Table but not in the numbers that played in those four days.

This now takes us to the unanswered questions.

Is there a way he could win like his Virginia friend and possibly more?

Can he win DC lottery every single day?

Is there really a solution to gaps when it comes to playing the lottery?

The answer to the above questions is an unequivocal 'Yes.'

He will need to master how to build and make use of Pro Tables.

He will master the language of numbers.

He will become a master in the core tools you need to win every lottery game.

These are the tools you will not find in an everyday lottery market.

If you are ready to join the few that win consistently, be ready to put in some work.

Everything worthwhile comes with some effort.

The process and information in this book is going to be heavy.

You will see me zero down on actual winning numbers consistently.

You will know how to have the winning numbers and when and why to play them for the winnings.

This book is not a novel and will not read like one.

It requires studying. You may need to read it several times. You can stop whenever you need to and resume when you are ready.

There is no lottery number that you cannot win.

If you are ready, let's go to work.

The Finished Table has not been favorable to DC friend.

His solution will be to use Pro Tables and the other necessary tools that will net more consistent winning for him.

I know that you are eager to see Pro Tables and how he would make use of them to enjoy more winnings.

I will put down Pro Table 1 down at this point, go into lottery number tools and come back to show you how a DC friend could indeed make as much and, possibly, more money than his Virginia friend.

Pro Table 1

	A	B	C	D	E	F	G	H	I	J
1R	1	0	4	0	8	3	5	5	5	9
2R	9	5	4	2	8	3	1	8	0	4
3R	0	6	0	6	4	9	7	2	4	9
4R	6	6	7	8	7	2	8	5	6	7
5R	2	1	2	3	9	3	3	7	1	1

Lottery Numbers

You have seen where DC friend had all the common two digits in a group and still could not win. The reason for that is because the table he was using had gaps. The gaps made it nearly impossible for him to enjoy similar winnings with his friend across the state line.

His Virginia friend is also going to have a similar problem at some point. The reason for that is because the groups in the Finished Table rotate and will continue to do so with the gaps.

You have to cure the gaps to enjoy consistent winnings.

You can indeed see the possibility of winning every single day when you study this book very well. Note that I did not say read. I mean it when I say that you must STUDY this book.

If you master how to close gaps and develop tables with zero inefficiency, you can indeed win every single day.

I will take you to the master level by the time you finish reading all the pages on Pick 3 and Pick 4.

Lottery Number Solutions come when you master the actual number movements.

The lottery numbers move in three different ways.

I will touch on the tools to help you capture consistent winnings.

You can read in great detail about the three methods from the number one lottery book in the world,

Lottery Little Book by Author Encoe

That is the only book where you will find the methods I am about to discuss.

I will bring the Pro Table 1 down at the end and show you how to apply these methods and enjoy more consistent winnings like the few elite out there.

Let the studying begin.

Lottery numbers move in three different ways.

Everybody out there understands numbers precisely the way they are taught in schools. People tend to look at the numbers in numeric order: 1, 2, 3, 4 and so forth.

The numeric numbers could be used to form trends. You need trends to win lottery games. There is indeed no other way lottery numbers play outside trends despite popular belief out there to the contrary.

You can form Pick 3 trends with the same numeric order like 123, 456, 789 and so forth.

The big issue for many is when the trend changes and appears like the following groups 439, 472, 652, etc.

When this happens you might see some people hit their head on the walls to challenge it. They are challenging the fact that those are actually trends for the simple reason that the groups did not appear in numerical order.

Those are numbers from columns A, B and C of the Finished Table. The Finished Table is of course formed from the same numeric order of which you participated in developing.

You need to also master how the numbers move along with Pro Table formations to begin to enjoy consistent winnings. You will know how and when to take advantage of the lottery games. You can apply that knowledge to every lottery result.

The future results are built into prior ones thereby giving you a huge advantage.

Lottery numbers move in three different ways.

You have the numbers that people are familiar with and use on a daily basis. The numbers, from the basic numeric order to more complex ones, are used daily in tasks as simples as counting your money to many other complex processes.

The basic numbers as I stated earlier are the ones in the numeric order like 0, 1, 2, 3, 4, 5, 6, 7, 8 and 9.

These are the numbers we used in forming the finished table.

The secondary movements of numbers are defined as Counterparts.

It is extremely important that you understand this part of the process.

You cannot hope to win lottery numbers every day without putting forth the effort to have all the necessary tools.

The secondary movement of numbers is called Counterpart.

A Counterpart System changes the numbers or trends by five (5) steps.

Here is an instance of a Counterpart application: you play numbers 123 and the lottery board plays 678. You can see that the two Pick 3 numbers are different. However, if you add 5 to each one of those original numbers, you will form 678.

        123

+   555

_____

        678

_____

You must always remember not to carry the remainder forward.

For instance, 789 plus 555 will be 234.

You cannot add 5 to the number 9 and call it 14. The number 9 plus 5 is equal to 4.

With this knowledge the Counterpart of 789 is 234.

You cannot win the lottery when Counterpart is applied to the number you are betting on unless you understand and apply it when necessary. Counterpart order is simple yet powerful.

Let me create one simple opportunity that DC friend would enjoy with the knowledge and application of the Counterpart System from one section of Finished Table that was not there when his friend was winning.

Finished Table

A	B	C	D	E
4	3	8	5	4
4	7	2	3	4
6	5	2	1	3
6	1	6	4	0
0	7	5	7	3

The above is just a section of the Finished Table for demonstrating the efficacy of the Counterpart system at this point.

If you look at the above table in the respective columns you will notice the Pick 3 numbers 303,147, 265, 517 and 660.

Let me change the same groups from columns E through A into the format most people will find easiest to read.

303

147

265

517

660

The above groups are the same Pick 3 numbers.

I will now apply Counterpart to the last digits. That means adding 5 to each of the last digits without carrying over.

303 plus 5 on the last digit 3 will become 8 thereby making it 308.

147 plus 5 on the last digit 7 will become 2 thereby making it 142.

The same process applies to each of the last digits of the groups.

The finished groups with Counterpart application on the last digits will go from the ones above to the ones below,

308

142

260

512

665

This, of course, is done by DC friend in the quest of trying to create winning numbers. You must always have the mindset of creating winning numbers. The lottery boards are not going to give them to you. It would cost them money.

Remember that they are in the business of making money, like every other business out there.

Now that DC friend altered the group through Counterpart application, let us see if he created success and took advantage of the DC 3 lottery game.

The winning DC 3 numbers from 6/16/12 through 6/19/12.

855	260
403	101

245   521

811   937

The trend on the same date started with winning DC 3 number 260 that played straight. They skipped 142 and played 403, of which 308 would produce a winning number on parlay betting. They came back and played 245, of which 142 would have produced another winning number on the parlay before completing it with 521, of which our method produced 512.

You can see that he was able to create four winning numbers in a span of three days from a trend where none existed prior to the application of Counterpart.

Let us take a look from another angle of the same table he used to create the opportunities.

Finished Table

A	B	C	D	E
4	3	8	5	4
4	7	2	3	4
6	5	2	1	3
6	1	6	4	0
0	7	5	7	3

You can see the winning number 521 in the third row of columns B, C and D.

The above table did not accord him any opportunity because it had gaps despite the fact that 521 is among the groups in the Finished Table.

The winning Pick 3 numbers directly above and below 521 (723 and 164) did not show up in the trend. He easily could have changed it by applying the Counterpart System based on the DC 3 results over the same period.

One instance will be by creating a Counterpart System through columns A, B and C.

Your goal must always be to create two common numbers before betting on the third one.

It may not play right below each one in every state but it will definitely put you in a position to enjoy more winnings.

If you apply the Counterpart in this format to the last digit of the first Pick 3 number 438, you will create 433.

Apply it to the middle digit of the next Pick 3 number 472, you will create 422 and lastly apply it to the first digit of the third Pick 3 number 652, you will create 152.

This is like applying the Counterpart in forward slash order.

The newly minted Pick 3 numbers will be,

433

422

152.

You might ask, "Why did you do that?"

I embarked on it once I noticed two common numbers that gave me the opportunity to create the third one.

The 'two common digits' rule was met with 433 for 403, 422 for 245 and profitable 152 for 521.

The Counterpart System gives you the opportunity to recreate and capture most of the trend and turn them into money.

You must remember that the Counterpart should be applied to any or all the digits as necessary to capture the winnings. I can create endless winnings from the same groups we have been discussing.

If you are looking at Pick 3 number 123 for instance, the Counterpart could turn the same number into,

623, 173, 128, 673, 178, 286, 678 and of course the same 123 that started the Counterpart application.

In the above instances I added 5 to the first digits, 5 to the first and middle digits and 5 to the entirety of the digits.

The winning Pick 3 number 123 could be turned into any of those.

Do not forget that you are not carrying over what remains after your addition. For instance, 9 plus 5 is equal to 4 instead of 14. The Counterpart of 5 will be: 5 plus 5 equals 0 (zero) instead of 10.

Once you master how to add the Counterpart, it becomes increasingly difficult for the winning trend to escape from your grip.

You noticed so far how I used the Counterpart to create winnings for our DC friend where none existed. We achieved that by just adding 5 into one digit of the respective groups that created the two common necessary digits.

You actually create more opportunities when you make use of adding as necessary to one, two and all the digits as the case may be.

You should make an effort to understand the Counterparts of any number that is in front of you.

Let me put down traditional numbers and their Counterparts right below them.

Numbers

0	1	2	3	4	5	6	7	8	9

Counterparts

5	6	7	8	9	0	1	2	3	4

As simple as the above numbers are, they will put serious money in your pocket once you master how to use them.

One quick way to study them, as I previously mentioned, is by adding the number 5 to the digits.

For example 123 will become 678, 223 will become 778 and 850 will become 305.

By now you should be entertaining the possibility of winning every single day when you do your homework.

We have so far discussed two ways lottery numbers move. In the above two, most of the winnings came through parlay betting. A good number of people may think that parlay is not enough for the simple fact that you win about $50 on a $1 bet.

If you master it and win just that on the midday and evening Pick 3 numbers it comes out to about $3000 every month. I am not even counting other straight and box winnings.

I played $10 on a 073 parlay that paid about $500. You should of course bet based on your mastery of the game and your budget flexibility.

I will now introduce the ultimate tool before going into the Pro Table.

This tool, in conjunction with the numbers and the Pro Table, will give you everything you will need to cash in on the winning numbers consistently.

If you are among the master class, continue reading.

Buckle your belt and get ready for a ride to some serious money.

Shadow System

You cannot expect the winning numbers to constantly play in numeric order. In fact they rarely do. You will often see winning numbers close to the last one that played from the table, and yet many cannot catch them.

You equally have gaps that move the winning numbers from one end of the table to the next. They move to the degree that is nearly impossible for many to catch.

Nevertheless, the fact remains that the winning numbers are related.

You can bring any and every group you are playing on into the winning circle when you master the three methods and steps to build a Pro Table.

I will discuss the third method, Shadow, after which I will bring down Pro Table 1 before proceeding.

I will show you at that point how you can win every lottery game in front of you. I must remind you once more that this will require some reading.

You will be amazed to find that it helps to read this book several times.

Keep in mind that most people put in an average of eight hours on their job in order to make a decent living.

If you are looking to make decent money from playing the lottery, be prepared to put in the work.

The strongest faculty you will ever possess is knowledge.

Once you acquire knowledge, no one can take it away from you.

Shadow Number System

The Shadow System is a bit more complicated that the Counterpart, in which you add 5 to every number.

You will understand the Shadow method by watching me put it into practice.

I will place Shadow Numbers against the traditional numeric system that you learn in schools.

Please make every effort to master the Shadow Numbers, they will come very handy in your quest to win every lottery game.

Numbers

0	1	2	3	4	5	6	7	8	9

Shadow

3	4	8	0	1	7	9	5	2	6

There is no lottery number that can elude you once you master Numbers, Counterpart and Shadow Systems, in conjunction with Pro Table construction.

Let us put the Numbers, Counterpart and Shadow digits below in that order.

Numbers	0	1	2	3	4	5	6	7	8	9

Counterpart	5	6	7	8	9	0	1	2	3	4

Shadow	7	9	5	2	6	3	4	8	0	1

I will change the above positions in the order of Numbers, Shadow and Counterpart.

Take note of the apparent changes.

Numbers	0	1	2	3	4	5	6	7	8	9

Shadow	3	4	8	0	1	7	9	5	2	6

Counterpart	8	9	3	5	6	2	4	0	7	1

I will now put the entire group in the order that lottery numbers move within one table.

The order will be Numbers, Shadow and Counterpart.

I will use the notation #, S and C to represent Numbers (#), Shadow (S) and Counterpart (C), in order to make the groups easier to identify.

Number, Shadow and Counterpart Groups

#	0	1	2	3	4	5	6	7	8	9
S	3	4	8	0	1	7	9	5	2	6
C	8	9	3	5	6	2	4	0	7	1
S	2	6	0	7	9	8	1	3	5	4
C	7	1	5	2	4	3	6	8	0	9
S	5		7	8		0		2	3	
C	0		2	3		5		7	8	

The above groups form the basis through which all lottery numbers move.

All lottery numbers will consistently rotate within the above groups.

The groups will consistently help you to capture the trends.

Let us take a quick look at trends under 0 and 1 to show you how to read it.

The entire group forms a circle and the numbers will consistently rotate within the circle.

If the trend you are following played 01 followed by 34, they are likely going to continue the trend with 89.

You will see numbers 34 right below 01 and 89 will be the next after 34.

If the lottery board plays 89, you will naturally expect the trend to continue by playing 26.

If, on the other hand, they play 13 instead of 26, you will know that the trend shifted to the column under 6 and 7. The 1 and 3 are in the same row with the 2 and 6 that you expected to drop.

What happened just now means that they are likely going to play 6 and 8 below the 1 and 3 that just played or the 4 and 0 above it.

You may recall that you get a Counterpart digit by adding 5 to the original number. If you do not recall, please go back and read the beginning of this chapter.

The trend we are looking at now originated from 0 and 1. The Counterpart of 0 and 1 is of course 5 and 6.

The Counterpart under the 0 and 1 produced 2 and 6 in the fourth row, after which you will get 7 and 1, which happen to be the Counterpart of 2 and 6. The rule is in order under the same trend.

The 6 and 7 at the beginning of the trend produce 1 and 3 in the same row with the 2 and 6 that resulted from 0 and 1 column.

If 34 is equal to 89 and 89 is equal to 26, that means that 95 is equal to 40 and 40 is equal to 13.

The same could and does apply to the other rows and columns.

It is very important that you do a careful study of the above table. If you read it several times, you will get to know them it by heart. That effort will help you tremendously in your lottery-betting journey.

Let me give you a hint of what you could accomplish with the above Number, Shadow, and Counterpart table in order to show you that it is the most powerful table when it comes to lottery games.

Let us look at the group under 123.

The Pick 3 groups under that trend are,

123

480

935

607

152

X78

X23

Now take a look at Virginia lottery Pick 3 results from 6/15/12 through 6/21/12.

The results on the dates are 532, 584, 853 and 207.

All of those numbers played in the evening drawing.

532 played on 6/15/12 skipped one day in each case.

All four Pick 3 numbers produced winnings based on our table.

532 (123), 584 (480), 853 (935), and 207 (607).

In each case the two common digits are together, thereby eliminating the trouble of worrying about the number in the middle that could split them. There is ample evidence here to give the bettor two winning numbers on parlay betting. The winning numbers played right after each other in the same way our method produced the groups.

Take a look at the Virginia Pick 3 midday results of 6/16/12, 6/17/12 and 6/18/12. You will see the winning numbers 917, 609 and 338. The three winning Pick 3 numbers would have equally produced winnings under the 012 group of our table.

The numbers would be 917 (715), 609 (260) and 338 (893). You can see that the group would have produced winners. You should strive to catch at least one of them. You will have the same opportunities on a consistent basis if you carefully study the table alongside your state lottery results.

There are times when the trend will be playing between midday and evening results.

Let us examine that with another state to show you why it is important to master the three ways that lottery numbers move, as well as how to apply them to your benefit.

You must pay close attention to identify the trends.

In the same time period the Maryland Pick 3 lottery played 531 on 6/13/12, 367 on 6/15/12, 852 on 6/17/12 and 770 on 6/19/12.

You will notice that the winning Pick 3 numbers played one day apart and all of them played on odd number dates, like the ones that played in Virginia.

The bettors in Maryland will win those drawings if they follow the trend under 678.

The numbers that played in Virginia were the groups under the 123 column.

The 678 is, by the way, a Counterpart of 123. That in itself is one reason why the two states played similar trends within the same time frame.

The winning Pick 3 numbers under the 678 column are,

678

952

407

135

680

X23

X78

Maryland started the trend by playing 531 (135), 367 (678), 852 (952) and 770 (407).

You can see that all the groups won like that of Virginia based on parlay betting.

The trend this time started by playing the first one during the midday drawing, followed by the next two that played in the evening drawing, and completed the trend by playing 770 for the midday.

These are very easy to recognize once you master the three ways lottery numbers move.

Take another close look at our table and how you can use it to capture more consistent winning lottery numbers.

I am going to use Virginia lottery Pick 3 winning results and put the numbers from our table beside them in brackets before proceeding to explain further.

I will be using the Pick 3 numbers from our table under column 234.

Virginia lottery Pick 3 played this trend in July 2012:

410 (801) for 7/1/12

651 (356) for 7/2/12

670 (079) for 7/3/12

205 (524) for 7/4/12

The above Pick 3 numbers will win on parlay betting. We are now graduating up to more winnings.

In the above groups you can clearly see the two common digits among them.

The newly introduced numbers in the Virginia Pick 3 winning numbers are 4 for 410, instead of 801 that our table produced.

I will put all the newly introduced Virginia digits down.

They are 4, 1, 6 and 0.

4 replaced 8, 1 replaced 3, 6 replaced 9 and 0 showed up for the last Pick 3 to replace 4.

If you look at the newly introduced numbers you will notice that the replacements followed the three methods by which lottery numbers move.

4 followed the Shadow 1 and 1 followed the Counterpart 6. The last number should be 9, which would be a Counterpart of 4 from 524, which would have made the fourth Pick 3 winning number 295 instead of 205 that played on 7/4/12.

They indeed played it as 952 exactly one month later to complete the trend; Virginia played it on 8/4/12.

This is another potent example, showing how the winning lottery numbers move.

Hopefully, you now realize that you can actually catch winning lottery numbers with the methods explained thus far.

There is much more to come, especially in the lottery Pick 4 section; You will be able to achieve unimaginable heights of success.

At this point, I will now bring down the Finished Table from which Pro Table 1 is carved out.

You completed the step by step construction of the Finished Table with me and we are now going to do the same in order to construct Pro Table 1.

I will delve into more explanations on how you can use these tables to enjoy more consistent winnings after constructing the Pro Table 1.

Finished Table.

A	B	C	D	E	F	G	H	I	J
4	3	8	5	4	8	2	0	2	5
4	7	2	3	4	5	9	1	6	1
6	5	2	1	3	8	6	7	0	1
6	1	6	4	0	9	8	7	2	9
0	7	5	7	3	9	0	3	8	9

The above is the table we started with. You will remember that our DC friend could not win like his friend in Virginia because the Finished Table had gaps and wasn't playing the trend in his market.

Even though DC 3 played 260 and 521 in the same time period, he was still unable to win. He did not even have the trends in place to give him a betting chance. The closest he came to winning was the 443 in the Finished Table but DC 3 played 403.

Our DC friend could not continue to wallow in that misery. He had to do something about it. At this point, I have shown you that he could make abundant winnings by applying the Counterpart and Shadow systems into the group.

Another thing he could do is reverse the trend and set it in his favor with the numbers available to him in the Finished Table. The result of that is the creation of Pro Table 1.

Pro Table 1

	A	B	C	D	E	F	G	H	I	J
1R	1	0	4	0	8	3	5	5	5	9
2R	9	5	4	2	8	3	1	8	0	4
3R	0	6	0	6	4	9	7	2	4	9
4R	6	6	7	8	7	2	8	5	6	7
5R	2	1	2	3	9	3	3	7	1	1

If you are not sure at this point as to its construction, I must ask you to go back and study the creation of the Finished Table. I will be completing the layout of Pro Table 1 on the assumption that you know how to build the Finished Table.

You will not get the luxury of a robust explanation of the Pro Table 1 construction.

You will, however, find it very easy and helpful when you become conversant with building the Finished Table.

With that being said, let us begin the construction of Pro Table 1.

You must remember that we built the Finished Table from numbers 0 through 9.

I start here by filling the grids with numbers from the Finished Table.

I began the process by using the columns where 260 and 521 appear, as they played in the DC 3 lottery. Those two were the numbers that played straight and our DC friend could not catch any of them.

I began by placing the numbers in column I of the Finished Table from column A, row 5R in a diagonal pattern like the one below.

	A	B	C	D	E	F	G	H	I	J
1R					8					
2R				2						
3R			0							
4R		6								
5R	2									

You can see that I started from the beginning of 260 in column I. I now continued the process by placing the numbers in column B of the Finished Table. The reason I chose column B is to capture number 5, which is the first digit of 521 that I mention earlier.

The B column numbers when placed in the grids will appear as follows,

	A	B	C	D	E	F	G	H	I	J
1R										
2R						3				
3R							7			
4R								5		7
5R									1	

The two constructed tables above will now appear as the one below

	A	B	C	D	E	F	G	H	I	J
1R					8					
2R				2		3				
3R			0				7			
4R		6						5		7
5R	2								1	

I have now taken care of columns that touched the 260 and 521. The other number that our DC friend could not catch was 403, because our Finished Table has 443 in column E.

You can see that 260 and 443 of the Finished Table are in the columns position, while the 521 is in row position. Since I am reversing their trend in my favor I chose the 260 that is in the column and a section of the 521 that is in a row. I will capture the rest of the numbers in rows including the 443 and 21 from 521, since I have already placed numbers that caught 5 from 521.

What I am doing in essence is redistributing the entire trend in my favor. The 443 will be picked in a row, thereby placing those numbers in different sections of the table, as well as 21 from 521.

The 443, when distributed to different sections, gives you the potential to catch the 403. The entire reconstruction gives you a heightened chance of capturing the 521 and 260 as well.

You now have added tools of Numbers, Shadow and Counterpart patterns to make you an even more formidable lottery player.

I will now continue the process of reconstructing the table by placing the rest of the numbers in the rows.

Remember that I did the initial two by taking the numbers from columns I and B.

I will now take out the two columns before placing the rest of the numbers to avoid any mistakes.

Next, I will take numbers from the rows in the table below to the new one being constructed.

Finished Table.

A	B	C	D	E	F	G	H	I	J
4		8	5	4	8	2	0		5
4		2	3	4	5	9	1		1
6		2	1	3	8	6	7		1
6		6	4	0	9	8	7		9
0		5	7	3	9	0	3		9

I have now eliminated the two groups that we have used already. If I am continuing from the first row for instance, the numbers I will use are 4, 8, 5, 4, 8, 2, 0 and 5. You will see those numbers in the first row. I will do the same for the rest of the group.

Let us now bring down our newly forming table and continue. As a matter of fact, we will continue from the numbers in the first row.

The two construction tables above will now appear as the one below

	A	B	C	D	E	F	G	H	I	J
1R					8					
2R				2		3*				
3R			0				7			
4R		6						5		7
5R	2								1	

The process will continue from column I, row 3R. I will equally put a star beside the number to show you where to start.

The two construction tables above will now appear as the one below

	A	B	C	D	E	F	G	H	I	J
1R					8		5			
2R			2		3			8		
3R			0				7		4*	
4R		6						5		7
5R	2								1	

The next set of numbers will continue from column E, row 3R. I will once more put a star beside it. The table will look like the one below,

The tables above will now appear as the one below

	A	B	C	D	E	F	G	H	I	J
1R					8		5			
2R			2		3			8		
3R			0		4*		7		4*	
4R		6		8				5		7
5R	2		2						1	

I will continue placing numbers to form the rest of the groups. The next numbers will be 0 and 5 that will complete the first row of Finished Table. I will skip 2 of column I because I used it already when I was placing the 260 group. This means that I will use 0 and 5 and then continue into the second row.

The next four numbers that I will place will be 0, 5, 4 and 2. I skipped 7 of the second row because I have already used it when I began constructing this table.

I will, for your reference, place a star where the next set will continue.

The above table will now look like the one below,

The construction tables above will now appear as the one below

	A	B	C	D	E	F	G	H	I	J
1R			4		8		5			
2R		5		2		3		8		
3R	0*		0		4*		7		4*	
4R		6		8				5		7
5R	2		2						1	

Now that we have filled in a good portion of the table, we will definitely jump through those that are already filled in to complete the grids.

I will put an 'X' where I will place the next set of numbers in order to guide you. I will place just four numbers before continuing through the rest. Remember that the next placements will continue from column F, row 4R.

Now take a look at the X when placed where the next set of numbers will go.

The construction tables above will now appear as the one below

	A	B	C	D	E	F	G	H	I	J
1R			4		8		5		X	
2R		5		2		3		8		X
3R	0*		0		4*		7		4*	
4R		6		8		X		5		7
5R	2		2				X		1	

I will now replace them with actual numbers from the Finished Table where we stopped. The last one stopped at number 4 of the second row. The next one will begin with the number 2 since we have already used 7.

The above table will appear as the one below when we replace the X with the numbers.

The construction tables above will now appear as the one below

	A	B	C	D	E	F	G	H	I	J
1R			4		8		5		5	
2R		5		2		3		8		4
3R	0*		0		4*		7		4*	
4R		6		8		2		5		7
5R	2		2				3		1	

You will continue constructing the table by filling up the two spaces in column E, row 5R as well as column A, row 1R to complete one half of the grid.

I will put X's in the remaining two spaces before replacing them.

The construction tables above will now appear as the one below

	A	B	C	D	E	F	G	H	I	J
1R	X		4		8		5		5	
2R		5		2		3		8		4
3R	0*		0		4*		7		4*	
4R		6		8		2		5		7
5R	2		2		X		3		1	

The construction tables above will now appear as the one below when the X's are replaced with the actual numbers

	A	B	C	D	E	F	G	H	I	J
1R	1		4		8		5		5	
2R		5		2		3		8		4
3R	0*		0		4*		7		4*	
4R		6		8		2		5		7
5R	2		2		9		3		1	

You have now completed one half of the table.

Now that you have completed one half of the table, let us fill in the rest of the grids with the remaining numbers from Finished Table.

You will begin from column J, row 5R. I will put double X's in the first six lines of the direction you will follow to fill in the numbers before beginning the process of completing the grid.

	A	B	C	D	E	F	G	H	I	J
1R	1		4		8	XX	5		5	
2R		5		2	XX	3	XX	8		4
3R	0*		0		4*		7	XX	4*	
4R		6		8		2		5	XX	7
5R	2		2		9		3		1	XX

I will now replace the XX's with the numbers from the Finished Table. You must remember to continue from where you stopped during completion of the first half of the table.

	A	B	C	D	E	F	G	H	I	J
1R	1		4		8	3	5		5	
2R		5		2	8	3	1	8		4
3R	0*		0		4*		7	2	4*	
4R		6		8		2		5	6	7
5R	2		2		9		3		1	1

In the above case I replaced the X's with the numbers from the Finished Table. I will now continue the process and complete the rest of the grids.

Your finished Pro Table 1 will look like the one below:

Pro Table 1

	A	B	C	D	E	F	G	H	I	J
1R	1	0	4	0	8	3	5	5	5	9
2R	9	5	4	2	8	3	1	8	0	4
3R	0	6	0	6	4	9	7	2	4	9
4R	6	6	7	8	7	2	8	5	6	7
5R	2	1	2	3	9	3	3	7	1	1

You have now completed the task of transforming the Finished Table into Pro Table 1.

I do believe it is safe for me to state that you now understand how to construct winning tables. This will be your first step to consistent winnings.

You can tweak the tables in different directions and still have winning trends.

We formed the Pro Table from the Finished Table that we constructed from the basic numeric order of numbers.

I will now show you Pro Table 2, which I constructed from Pro Table 1.

At this point I will not assume that there is the need to explain the methods, since we have been dealing with them in-depth up to this point.

The only thing you will need to do now is be able to tweak the tables according to your imagination.

The trends will still be intact.

The next method you will learn will be in the Pick 4 chapter; be certain to read the section on Pick 4.

Pro Table 2

	A	B	C	D	E	F	G	H	I	J
1R	6	7	6	9	8	9	5	1	4	3
2R	5	5	4	7	8	8	2	0	3	1
3R	2	9	6	4	0	7	6	0	4	9
4R	8	6	3	8	5	7	7	0	4	9
5R	1	2	3	3	0	1	2	2	5	1

Catching the Winning Numbers

The same trend works in every lottery market.

You can also identify winning Pick 3 and other lotto games with the table.

Pro Table 3

	A	B	C	D	E	F	G	H	I	J
1R	0	4	4	3	2	1	9	8	8	1
2R	3	6	7	1	4	3	2	7	0	5
3R	9	2	0	7	6	0	4	7	6	9
4R	2	5	1	7	6	9	8	5	6	7
5R	4	9	8	8	5	5	4	5	3	6

Pro Table 4

	A	B	C	D	E	F	G	H	I	J
1R	6	4	0	2	2	9	7	4	0	5
2R	0	7	9	3	6	1	7	6	5	2
3R	2	6	4	7	1	6	9	8	3	3
4R	7	4	1	4	5	5	8	8	5	1
5R	0	5	3	8	7	9	9	4	8	6

Pro Table 5

	A	B	C	D	E	F	G	H	I	J
1R	4	3	6	0	1	4	1	7	1	4
2R	1	3	8	8	2	6	3	4	0	5
3R	7	6	3	4	5	1	1	1	7	3
4R	1	9	2	5	5	7	0	2	1	5
5R	8	2	2	4	2	5	0	5	4	2

Pro Table 6

	A	B	C	D	E	F	G	H	I	J
1R	6	8	3	2	0	8	8	4	1	8
2R	9	8	9	7	6	6	3	7	8	2
3R	5	3	7	3	4	5	4	8	6	2
4R	9	5	2	2	9	4	1	6	0	7
5R	1	7	7	1	1	5	9	6	5	2

Application of Number, Shadow and Counterpart methods will put you outside the realms of lottery betting. The level of control you will develop in catching the winning numbers is unheard of. I have so far shown you how easy it is to capture two numbers on most Pick 3 games and take advantage of it through parlay betting.

Let us now look at getting consistent Pick 3 numbers based on the three methods with Pro Table 3 and Virginia Pick 3 lottery past lottery results.

Precision Betting

Pro Table 3

	A	B	C	D	E	F	G	H	I	J
1R	0	4	4	3	2	1	9	8	8	1
2R	3	6	7	1	4	3	2	7	0	5
3R	9	2	0	7	6	0	4	7	6	9
4R	2	5	1	7	6	9	8	5	6	7
5R	4	9	8	8	5	5	4	5	3	6

Virginia lottery Pick 3 played 467 midday and 642 evening on 8/1/12 to begin that month.

You will find 476 in Row 3R, column G, H and I. Above the 467 on the same table you will see 270 and 988. You will be amazed to notice that Virginia played 898 on 8/4/12 and 072 on 8/7/12.

The three Pick 3 numbers that won played in precisely three days apart. Go to Virginia lottery and check out the Pick 3 numbers that I mentioned.

The evening Pick 3 numbers that played on 8/1/12 was 642.

In Pro Table 3, you will notice that there is no 642. This is when the three methods come handy. You will embark on creating 642 based on the three methods. The goal is always to work out two prior results to enable you create the trend and continue with the third winning number.

You can create more 642 from all possible sections of Pro Table 3 and identify where the trend formed before betting on the numbers. You must bear in mind that the 642 is not necessarily in the same trend with 467 that played during the midday result. They could depend on the table you are creating the trend from.

I began the process by looking through the numbers in Row 1R first. I will do the same thing with all the rows and columns of the Pro Table.

I noticed that I could form 642 from the Pick 3 number 219. You will see 219 under column E, F and G. I will now put down all the Pick 3 numbers under the same column.

The entire group will be like the one below

E	F	G
2	1	9
4	3	2
6	0	4
6	9	8
5	5	4

If I apply Counterpart to number 1 of the 219 it will become 6. Remember that Counterpart adds 5 without carrying the remainder as I explained earlier on. If you add Counterpart to 9 of 219 it will become 4.

If you leave the lead number 2 untouched and add Counterpart to the 1 and 9 you will get 264. The 264 is the same number Virginia lottery played as 642 on 8/1/12.

I have to equally change the rest of the group for the trend to remain intact.

The new group with the application of the Counterpart on the last two digits will go from the one above to the new one below

E	F	G
2	6	4
4	8	7
6	5	9
6	4	3
5	0	9

Now that we have created the 642, we will then proceed to check the new trend against actual Virginia result. Bear in mind that if this did not accomplish what I am looking for, I will continue with the other ones through which I will capture the trend they are playing.

You will notice that Virginia Pick 3 played 847 on 7/30/12 followed by 059 to end the month on 7/31/12 and finally the 642 that started the new month on 8/1/12.

The three winning Pick 3 numbers played just one day apart. You should be able to catch the 642 at least.

The Pro Table 3 is so efficient that you can go to any state of your choice, recreate any past winning number and capture the rest of the winning numbers in that trend within a relatively short period of time.

You have just witnessed how I zeroed in with precision on actual winning Pick 3 numbers in the trends by employing Shadow and Counterparts with the Numbers.

Pro Table 4 below is even more powerful than Pro Table 3.

I created Pro Table 4 from Pro Table 3 by using the table construction I discussed in the earlier chapters. I do highly advice you to trace and recreate Pro Table 4.

If you place Pro Table 3 and 4 side by side and take a careful look, you will begin to actually see how I created Pro Table 4 from 3.

You can do it by using the same method I discussed in creating the Finished Table.

Pro Table 4

	A	B	C	D	E	F	G	H	I	J
1R	6	4	0	2	2	9	7	4	0	5
2R	0	7	9	3	6	1	7	6	5	2
3R	2	6	4	7	1	6	9	8	3	3
4R	7	4	1	4	5	5	8	8	5	1
5R	0	5	3	8	7	9	9	4	8	6

Go back to Virginia past Pick 3 results against Pro Table 4 above.

Virginia lottery Pick 3 played 937 on 7/31/12 followed by 467 on 8/1/12 and 144 on 8/2/12. The winning numbers once again played right after each other in a span of three days.

You will find all three winning Pick 3 numbers in column B, C and D of Pro Table 4 above. They are in 2nd, 3rd and 4th rows of columns B, C and D of Pro Table 4 above.

I have so far shown you different ways of getting the winning numbers. You can do the same thing every single day through every month.

The key is mastering how to build the tables, do careful study of the trends and create one where it does not exist. I can recreate and walk you through every winning Pick 3 number from just one of the tables. I create different tables; apply the three methods when necessary until I zero in on the winning number.

In your quest of trying to recreate two winning Pick 3 numbers to get the trend, do not forget to go back through the last two weeks to make sure that one of the winning numbers you have just created had not played.

If any of the numbers you just created played already, it is another confirmation that your trend is good.

One instance of such will be from the first, second and third row of columns D, E and F of Pro Table 4.

You will see the winning numbers 229, 361 and 716 below each other. If you minus one (1) from each of the last digits of the group you will form the following Pick 3 numbers 228, 360 and 715 from the group. You will notice that Virginia Pick 3 played 715 three days earlier.

Once they continue by playing 360 on 8/2/12 you will immediately continue by playing 228 which played the next day.

If I am betting on the game I will play both 715 and 228 for the next few days.

Most trends will play three in the group and may move on to another opportunity. They will come back at some point to play the rest of the groups.

One other important thing to bear in mind is that you could still recreate the new trend or capture the other ones with application of Counterpart and Shadow.

There are instances when they will minus or add one to one of the groups before playing the other two.

An instance of that will be if the Pick 3 group that is being played are 123, 456 and 789 for example.

They may decide to alter the 789 as 689, 779, 788, 889, and so forth. In each of the Pick 3 examples above, you will notice that in each case one digit is different by addition or subtraction of one.

The group could play as 123 followed by 788 for instance. You will naturally not notice that the 788 could indeed be playing the role of 789.

The group of Pick 3 numbers like 123, 456 and 789 fall under what we call Difference One in lottery jargon.

The fact that 788 is introduced at this point does not mean that 789 is not going to play at some point. The lottery people are not in the business of making it very easy for you.

If they play 123 followed by 456, most people will likely play 789 or something similar. The bettors who played 789 and bet parlay as well will win. You should expect a lot of people to win.

If on the other hand they played 123 followed by 788, most people will not think of playing 456.

Let us look at actual example through Pro Table 4 and DC lottery.

I will be using the same time period like that of Virginia.

Remember that you can do the same thing with any state lottery of your choice. You will arrive at similar result regardless.

DC lottery played 472 and 129 on 8/1/12 followed by 667 on 8/2/12.

You would be looking to where you can recreate 472 and 129 that just started that month. You need to have Pick 3 numbers that would represent the two that played to be able to have trend that will enable you bet.

Pro Table 4 would give you the opportunity to make some money by going to column E, F and G. In the first three rows you will find the following Pick 3 numbers, 297, 617 and 169.

If you apply Counterpart to the middle digits (remember Counterpart is by adding 5), the Pick 3 trend will change from 297, 617 and 169 to the newly formed group of 247, 667 and 119.

You will notice that DC lottery played 472 followed by 129.

In the conversion 472 is given and 129 is very close to the 119 that I just recreated. I will definitely embark on playing 667. I will of course play the 667 as well as bet parlay on it. The primary reason why I would consider betting parlay is because I had 119 against 129 that they played.

I will consider that prudent in case they altered one of the numbers from 667. In this case the 667 played straight. If you bet $1 straight and box, you will end up getting about $660.

If you equally added parlay betting in this case you will end up getting about $50 for $1 on 66X as well as another $50 on X67.

This game would cost you about $5 and pay you about $760. That would be considered good money on $5 wager by any standard.

If you master the three methods that lottery numbers move, and apply them properly, you will consistently create and enjoy winnings from trends that otherwise would not exist.

Let me demonstrate this with example from Maryland lottery Pick 3. I am going to use the Pro Table 4 and do it in the same time period.

Please pay close attention because I will go a little deeper here. Lottery numbers are much more than what people think. You will know a lot more if you are patient enough to study the Pick 4 section. This is just a tiny piece of what is yet to come.

Maryland lottery evening Pick 3 results played 090 on 8/1/12, 337 on 8/2/12, 917 on 8/3/12 and 581 on 8/4/12. I will put the above winning Pick 3 numbers right below each one in their order to help in the explanations.

090

337

917

581

I will recreate the above trend from Pro Table 4, columns D, E and F. In those columns you will see the following Pick 3 numbers. You will notice that there is no resemblance between the above results and the Pick 3 numbers from Pro Table 4

Pro Table 4

D	E	F
2	2	9
3	6	1
7	1	6
4	5	5
8	7	9

I will begin the process by converting the first two digits. The reason I am doing that is because I started with the initial Pick 3 result of 090. In the case of 229, there is already the number 9. I will convert 22 of 229 into 00 to create 009. I will do the same with the rest of the groups. Once I create 009 in this case and convert the rest of the groups, the entire trend will still be in place.

The order of conversion will be Numbers, Counterpart and Shadow

I will bring down at this point the two digits from the group that will be converted.

22	77	55	00
36	81	24	79
71	26	89	34
45	90	63	81
87	32	08	53

As you can see I converted the 22 into Counterpart (77), Shadow (55) and Number (00). I did the same process for the rest of the group.

The new two digits I will be using at this point to create the new group will be,

00

79

34

81

53

I will complete the sets of Pick 3 numbers in the group with the numbers in column F of Pro Table 4.

The new result will now go from

D	E	F
2	2	9
3	6	1
7	1	6
4	5	5
8	7	9

To the new groups,

009

791

346

815

539

I will now put the above new groups against Maryland actual lottery results.

Maryland lottery evening Pick 3 results played 090 on 8/1/12, 337 on 8/2/12, 917 on 8/3/12 and 581 on 8/4/12. I will put the above winning Pick 3 numbers right below each one in their order to help in the explanations.

090

337

917

581

You will notice on close observation that this trend actually started on 7/10/12 by Maryland playing 539 straight. The 539 is of course very convenient for the lottery bettor to forget. I am putting major emphasis on the word convenient.

Always remember to check if the trend you worked out started earlier. This will happen sometimes. The numbers are not meant to be very easy because it could become expensive.

Maryland played 090 per our workout followed by 337. The winning Pick 3 number 337 played as decoy. You will notice that our conversion produced 346. The winning Pick 3 number 346 is actually the same as 337. The sum total of 337 is 13 and the sum total of 346 is equally 13. If you had observed the 539 in the work out you will immediately know that they are playing the trend.

If you take 1 (one) from 4 of the 346 and add it to the last digit 6 of the 346, it will become 7, thereby making the number 337.

The next winning Pick 3 number that played was 917 followed by 581. On keen observation you could easily win two games from this group.

The more important point here is that you are able to create trends that captured the numbers they are playing.

You have endless trends from the tables and that is why it is important that you study the book thoroughly.

You can do this on just about every number that plays and zero in on the trend. You have more than ample ways of enjoying consistent winnings if you carefully study this book.

I highly encourage you to do this exercise on two different states. If you make mistake accept fault, go back, study the mistakes and make the necessary corrections.

You cannot look forward to cashing $500 on $1 bet consistently without studying the book.

It is important that you master different ways of looking at lottery numbers. If I tell you that all numbers 0 through 9 could fit in just two numbers with the application of Shadow and Counterpart, a lot of people will find it near impossible.

That reaction is of course understandable because people are not trained to know that.

If you truly study this book, you will be able to consistently zero in on any winning trend you are working on. You can work out the winning numbers with just two results. This of course is possible because you can represent all numbers with just 1 and 2.

I will demonstrate an example with actual Maryland lottery.

Let us stay in the same month of August.

At this point the only result available is 604 and 090 that played on 8/1/12. I am not going to use the benefit of looking at past results that will make it easier.

We are trying to use the two winning Pick 3 numbers and create trend that will give us additional winning Pick 3 numbers.

You will start by looking at all the tables to figure out which ones will give you the best entry point.

Please bear in mind that the same result could be achieved with any of the tables. You can do that with more than one or two sections of any of the tables.

The tables by themselves are actual winning trends. You can check them against any state or country past lottery results of your choice.

There are enough people that will not be willing to put the necessary effort to achieve the results.

The energy here is for those that are willing to study this material and prove the skeptics wrong.

Let me bring down Pro Table 5 and embark on creating winning trends based on the only available Maryland Pick 3 results as at the time (604 and 090 of 8/1/12).

	A	B	C	D	E	F	G	H	I	J
1R	4	3	6	0	1	4	1	7	1	4
2R	1	3	8	8	2	6	3	4	0	5
3R	7	6	3	4	5	1	1	1	7	3
4R	1	9	2	5	5	7	0	2	1	5
5R	8	2	2	4	2	5	0	5	4	2

After careful study of the above table, I decided to go with the winning Pick 3 numbers in columns A, B and C. I will bring that section down and begin the explanation afterwards.

	A	B	C
1R	4	3	6
2R	1	3	8
3R	7	6	3
4R	1	9	2
5R	8	2	2

In order to create the 604 that started the trend I will convert the middle numbers into Shadow that will now change the above groups after Shadow application into,

406

108

793

162

882

If you reverse the groups, the above Pick 3 numbers will now become the ones below,

604

801

397

261

288

You can work on it without reversing them if you chose to do so. You might be looking for the second winning Pick 3 number 090 that played after 604.

In the above case the 801 satisfied my requirement of 090. If you sum up 801 the result will be 9 or 090 as the case may be. I made decision to use the group as my trend to begin the month. I will be expecting them to play 397 based on the trend or go the other direction and play 288.

They continued the trend with one slight problem. Instead of playing the 288 per our calculation, Maryland played the Counterpart 337. I am now left with contemplating on changing the rest of the group into Counterpart or figure out what could be wrong.

You would in any case win parlay with the 397. You would win about $50 on the back pair or parlay X37 from the 397.

I highly advocate betting parlay along with the other numbers. The parlay, however, is not enough for some readers. I knew that the trend was right but not catching the entire group.

Should I abandon it and look for another opportunity?

No, because this workout is based on the fact that this is the only information available to us.

I will try to convert the winning number 801 into 090 to see if it will help.

How do I do that?

I will recalibrate the entire group with the mindset of creating 090. Remember that 604 is already given since we started building the trend with 604.

Let me bring the new group down before I start the new conversion.

604

801

397

261

288

In the last digits I can convert the number 1 of 801 into number 9. Let us convert the last digits of the group from,

4, 1, 7, 1 and 8 into Shadow that will give us the following results,

1, 4, 5, 4 and 2. I will now convert the new digits into Counterpart to create the required number 9.

6, 9, 0, 9 and 7. Remember that you get the Counterpart by adding 5 to each number.

I will now replace the last digits of the above group with the newly formed numbers that will now change the groups from,

604

801

397

261

288

Into the newly formed groups when the last digits are replaced,

606

809

390

269

287

I have now created 9 in my quest of trying to create 090. In the above case I have 06 from 606 as well as 09 from 809 in the first two winning Pick 3 numbers above. I am satisfied with the last two digits of the groups at this point.

I will now begin to convert the numbers in the first group to create zero (0) thereby achieving my goal of creating 090. I would hope that I can have the given winning Pick 3 number 604. I am not particularly perturbed about the 604 since that number already played to begin the trend. It will be nice to have 604 but not nearly as important as 090 instead of 801 that misaligned the trend up to this point.

Let me bring down the first digits before converting them to hopefully create 0 that will enable me create the number 090.

6, 8, 3, 2 and 2 into Counterpart will become 1, 3, 8, 7 and 7.

1, 3, 8, 7 and 7 into Shadow will become 4, 0, 2, 5 and 5.

I will now replace the first digits so that I will have 0 in place of 8 to achieve my goal of creating 090.

The new groups will now go from,

606

809

390

269

287

Into the one below when the newly formed numbers are applied,

406

009

290

569

587

As you can see the process recreated the 604 as well as 090. I have now converted the initial group with 801 into the current group with 090. This means that I drastically increased potential of winning more. You would end up winning parlay on the 049 with 290 that played.

The 569 played one week later on 8/9/12.

You would equally win parlay on 587 with the winning Pick 3 number 581that played on 8/4/12. The 581 that played on 8/4/12 actually gave you clue of when the remaining winning Pick 3 number of the entire group (569) will play.

You may wonder how I know. The dates have relationship with the numbers as well. The Counterpart of 4 of the winning date 8/4/12 for 581 is 9 of the winning date 8/9/12 for 569.

Let me take it a bit further without getting into the advanced stage.

The winning Pick 3 numbers that played as parlay: 290 and 587 have relationship in the same manner as the 049 of 8/2/12 and 581 of 8/4/12.

The Counterpart of first (lead) digit of 2 from 290 is 7 which is the third (last) digit of 587.

If you replace the lead number 2 of 290 as well as the last number 7 of 587 with Shadow instead of Counterparts you will get 4 in place of 2 thereby creating the winning number 490 that played as 049 which will in turn replace 7 of 587 with the Shadow of 4 which is one (1) thereby creating 581.

The information could be heavy. They are indeed heavy and should be studied over a period of time.

Go back and read Lottery Icon several times. You will be amazed to know that each time it will become clearer.

Number, Shadow and Counterpart Groups

#	0	1	2	3	4	5	6	7	8	9
S	3	4	8	0	1	7	9	5	2	6
C	8	9	3	5	6	2	4	0	7	1
S	2	6	0	7	9	8	1	3	5	4
C	7	1	5	2	4	3	6	8	0	9
S	5		7	8		0		2	3	
C	0		2	3		5		7	8	

Super Table 1

	A	B	C	D	E	F	G	H	I
1R	7	6	2	4	5	7	6	1	3
2R	8	5	8	9	9	3	3	1	3
3R	5	9	1	0	1	5	2	4	4
4R	0	2	5	7	8	2	6	9	3
5R	0	0	0	7	0	6	7	4	3
6R	5	8	2	8	0	8	2	9	1
7R	8	7	3	4	7	5	6	2	0

Super Table is the bridge between all the lottery games. If you fail to establish the trend you are chasing with all the Finished and Pro Tables, this is the place. There is no trend that can escape the efficacy of the Super Tables.

You may equally decide to use one or all the tables until you zero in on the winning game you are playing. I personally make use of all the tables. I have reached a point where if I do not see the trend I am looking for, I will immediately create one that will get me to that particular trend.

I have established through all the examples in this book that you can work out any trend from the tables.

There are patient bettors who prefer to play the same set of numbers.

Let me give you one tiny demonstration that those players as well as you could use from Super Table 1.

I will use the Pick 3 numbers from columns G, H and I of the Super Table 1.

Let me bring that section down before beginning the demonstration.

Super Table 1

G	H	I
6	1	3
3	1	3
2	4	4
6	9	3
7	4	3
2	9	1
6	2	0

You can convert the numbers in the group to create several scenarios all of which will still be intact as a trend within the conversion that you created.

This in essence means that the newly created group will and do work as a group. That group gives the patient bettors the opportunity to win more.

The above statement is fact because you will be choosing the group based on the trend that is playing in the first place.

Now let me convert the numbers in the Super Table 1, columns G, H and I above.

I will change the entire group into Shadow which will turn the above groups into the Pick 3 numbers below,

940

040

811

960

510

864

983

I have now completed the first process. The Pick 3 numbers from the above table as well as the newly converted ones are solid. They could be use separately or together to capture the winning numbers from the trend.

I created the new table to show you that you can go beyond the table at hand. Now let me take the new table above back to the conversion I really wanted to create for the purpose of this demonstration.

I will do that by leaving the numbers in the first digits from Super Table 1, columns G, H and I intact. Those first digits will form new set of Pick 3 numbers in the trend when combined with the second and third digits from the above formed table.

The new set of Pick 3 numbers in the trend will now look like the one below,

640

340

211

660

710

264

683

As you can see, I have now created three different trends from the G, H, and I column of Super Table 1.

Let us put actual winning Maryland Pick 3 numbers dates in the same period of time against the above formed new groups to see the efficacy of the Super Tables.

640 played in MD on 8/1/12

340 played in MD on 8/7/12

211 played in MD on 8/20/12

660 played in MD on 8/26/12

710 played in MD on 9/12/12

264 played in MD on 8/26/12

683 played in MD on 8/25/12

As you can see from the above newly formed trend, over 85% of the Pick 3 numbers played in that month of August.

The 710 completed the trend in less than one month. The numbers here created and will continue to create huge opportunities for the patient bettor and those who studied and apply the methods in this book thoroughly.

When looking at actual results give it some time. You could take a good one minute to study the results and try to find out why that number played. You can recreate why a particular Pick 3 number played. The answer will be apparent if you do it right.

You can actually pick additional opportunities when you put some effort in studying it.

Take a quick look at the above winning Pick 3 numbers that I just worked out with the actual dates. If you go back to Maryland lottery Pick 3 of those dates, you will notice that the numbers played on the dates beside them.

There is, however, additional opportunity that many will not see unless they look closely. Take another look at one of the Pick 3 numbers among the group, winning Pick 3 numbers 264.

You will notice that 264 played several times in that month. This means that the other winning Pick 3 numbers in the group will do the same when it gets to their respective turns.

A further look at 264 will show you that it played as 462 two days prior to the winning Pick 3 number 340.

It came back and played as 426 two days before 112 in the same manner it did earlier.

It came back and played two days prior to the winning Pick 3 number 606.

It played again as 642 two days after the winning Pick 3 number 368.

You can see that 264 created more opportunities that many would not see.

One of the mistakes many people make is by not being patient. It is very important that you see why and how the lotto number played.

You need the above tool to become a maestro.

You can employ the efficacy of the Super Tables in Pick 3, Pick 4 and every other lotto game for that matter.

There are endless ways of catching the winning Pick 3 numbers.

Pick 3 Galore

Take a very hard and patient look at the Pick 3 galore against the trend that your market is playing to make sure that you extract all the opulence.

339	359	321	218	253
722	622	866	649	421
198	229	409	746	391
348	721	810	263	257
622	198	340	811	820
548	644	404	926	858
339	220	403	348	104
226	920	216	404	630
640	448	026	513	147
348	578	359	217	337
138	986	650	725	461
956	147	382	409	581
613	138	219	129	091
986	856	875	202	955
036	602	522	394	685
636	138	511	416	046
096	827	793	497	176
989	232	260	488	162
606	114	711	350	497
372	023	489	716	239

Chapter 3

Pick 4 Lottery Book

I thought that I had won my share of Pick 4 lottery numbers until I decided to go outside the walls in recent times. I have never really settled for the axiom of odds of winning Pick 4 being 1:10000.

I am not, however, going to spend my time contesting it. I will leave that field to the crowd and concentrate on how to tilt the game to my advantage. That is precisely why I did not stop working on getting more Pick 4 winning numbers over the years.

I will take you to a new place and prove to you that you can indeed win more than you have ever imagined possible.

I have always enjoyed working on Pick 4 with the idea of breaking the 1:10000 odd convention.

I did this over several years.

The other day I was going somewhere and noticed a vehicle with a unique tag number. I decided to put the tag number in the new method I developed to see what happens. It led me to the winning Pick 4 number 2424 in Maryland.

The winning Pick 4 number 2424 came very handy.

The next day, I visited somebody and decided to apply the name of the facility (Sinai Hospital) in the new method. It produced the next winning Pick 4 number.

I decided to go further by employing a completely different language (not English) in the equation.

Guess what happened?

It produced another solid winning Pick 4 number.

This good fortune occurred right after each other.

If any group of 1000 people are taught the same theory over a period of time they will master that theory. If they go out and teach sets of one thousand people the same thing they learnt, the new group will in turn master the same theory.

If you test some people from different countries that had mastered the theory, they are likely going to give you similar answers because they studied the same thing.

The answers are precisely going to be the outcome that the developer of that theory came up with.

The development at this point becomes convention.

The convention works for the person or persons who developed the theory in the first place.

This must be one of the reasons somebody said the other day that nearly everybody go through the same BRAIN WASH.

As good as any convention is, the fact remains that there will always be room for more. If convention is not questioned and improved on as necessary, the masters of that theory will in essence operate in the world of mechanics.

They will always produce the expected results.

The results in this case do not often favor the bettor.

The lottery odds are tilted at two to one (2X1) advantage at best to the lottery owners. This scenario is of course based on when you win. If you factor the times that you lost the bet which of course is very often, the odds are worst than the mathematical odds.

You bet $10000 on Pick 4 lottery to win $5000 at best. This of course is based on the mercy that you did not miss or make any mistake on any of the numbers.

How do you capture the same type of Pick 4 winnings that I enjoyed recently?

How do you win Pick 4 consistently?

How do you decimate the conventional odds of 1:10000?

If you are ready, let's get to work.

The required numbers that form Pick 4 winning numbers are numbers zero (0) through nine (9).

The odds in itself are based on the probability of four numbers from 0 through 9 coming out.

Nearly every table we have formed started from the numeric 0123 and so forth.

If you look at numbers and letters, the only two that resemble are O in letters and 0 (zero) in numbers.

The other letter and number that will come close are 1 (one) in numbers and I (i) in letters.

The I in letters apply when it is used as capital letter of course.

If I am forming tables it will be difficult for you to figure out if I am using the letter I (i) or the number 1 (one)

I will begin basic table formation that will include O. You can decide on your own weather to apply it as O or zero.

In the above instances you are the only person who knows if you are using numbers or letters.

If on the other hand you apply both letters and numbers, it becomes increasingly difficult for the lottery house to know what you are working on.

Stay alert and pay close attention.

The lottery odd calculations are based on the fact that everybody follows the convention of 0123 order and so forth.

It will become increasingly difficult for the computer to understand the number you are working on based on the new table construction that I am going to demonstrate.

If you achieve this level of number trend mastery you would end up flipping the table.

One of the single most important reasons people lose is because they do not have clue on what the lottery house is going to play which is based on the numeric convention.

If on the other hand the lottery people cannot figure out how your trend is constructed which is not based on convention, you would in essence turn the table in your favor.

You will learn how to form unconventional tables later.

Unconventional Table Formation

One

	A	B	C	D	E	F	G	H	I	J	K	L
R1		0			9		6	7	8		5	
R2		1		0		8	5		9	6		4
R3	5	2	6	1		7	4		0	7	2	3
R4		3			2		3		1	8		
R5		4								9	0	1

Unconventional Table 1

One

	A	B	C	D	E	F	G	H	I	J	K	L
R1	0	0	2	5	9	9	6	7	8	9	5	4
R2	1	1	4	0	8	8	5	6	9	6	3	4
R3	5	2	6	1	2	7	4	8	0	7	2	3
R4	3	3	7	1	2	5	3	1	1	8	6	7
R5	6	4	0	3	4	7	0	2	5	9	0	1

Besides using the above table to identify winning numbers, you can exponentially increase your winnings by mastering how to create the winning numbers consistently.

I will use the above table against actual Maryland lottery winning Pick 4 number.

Let us go to middle of September, 2012 and see if we can actually recreate winning numbers from the above table.

Maryland played 0574 on 9/15/2012.

In Unconventional Table 1 above I can create 0574 from the winning Pick 4 number 0259. You will find 0259 in row R1 of columns B, C, D and E.

I will convert it along with the rest of the group under that column.

Let us bring that section down before embarking on the conversion.

B	C	D	E
0	2	5	9
1	4	0	8
2	6	1	2
3	7	1	2
4	0	3	4

The numbers under column B stays intact because 0 (zero) is the lead number for 0574 that we are trying to create.

I will convert the numbers under column C into Counterpart and then reconvert the newly formed number into Shadow to create number 5. The process will be as follows:- 2 is the original number converted to Counterpart that would produce 7. The new number 7 will then be converted to Shadow to produce number 5.

In this case you will have the lead number 0 plus the second digit 5 in your quest to produce 0574. Do not forget that you will do the same thing to the rest of the group under C column.

The entire group under column C after the application will now go from,

2, 4, 6, 7 and 0 when changed to Counterpart will be,

7, 9, 1, 2 and 5 which will finally change after Shadow application to,

5, 6, 4, 8 and 7.

As you can see I have now converted the entire group under column C.

Since I have achieved my goal of 0 and 5 for the lead and second digits I will convert the third digit under column D. The goal is to create the number 7 towards the task of getting 0574.

The numbers in the D column will go from,

5, 0, 1, 1 and 3 into the newly formed group below,

7, 3, 4, 4 and 0.

In the above case I only needed to apply the Shadow to achieve the required result.

The final task will be to get the last digit (4) from column E.

I can achieve this by converting the groups into Counterpart. Remember that you only add 5 to create the Counterpart.

The new set of numbers after the conversion will go from,

9, 8, 2, 2 and 4 into the new group below,

4, 3, 7, 7 and 9.

I have now converted the entire group.

The entire group after they are converted will go from,

B	C	D	E
0	2	5	9
1	4	0	8
2	6	1	2
3	7	1	2
4	0	3	4

Into the newly converted group below,

B	C	D	E
0	5	7	4
1	6	3	3
2	4	4	7
3	8	4	7
4	7	0	9

I will write down the same winning numbers without the table before I go into further explanations.

0574

1633

2447

3847

4709

You can see the entire group with the 0574 trend. The group moves in tandem and exhibit similar behavior. Lottery number trend stays intact when the trend is not violated.

Now that I have converted the group, let us check it against actual Maryland lottery results over that period of time.

I will put our worked out numbers side by side with actual Maryland Pick 4 results and the dates they played.

0574 MD played 0574 on 9/15/12

1633 MD played 1336 on 9/24/12

2447 MD played 4274 on 9/24/12

3847 MD played 8743 on 8/31/12

4709 MD played 0497 on 9/28/12

You can see that the winning Pick 4 numbers played over the same period.

The clear relationship is apparent because they belong to the same trend. The 1633 played more than once. Take a good look at the other dates it played with the other accompanying winning numbers including 7942 of 9/31/12 that is actually a decoy of 3847.

This level of mastery of number trend is possible because I used one hundred percent efficient table. You are now beginning to understand manual trend calculations to a level that is unheard of.

If you study this book per my instruction and follow it, you will be unbeatable.

In case you may be wondering if the above trend is just by mere luck, I will demonstrate one more with Unconventional Table 2 below.

Unconventional Table 2

	A	B	C	D	E	F	G	H	I	J	K	L
R1	7	8	5	3	9	2	4	8	1	0	1	2
R2	4	6	2	5	1	9	1	5	4	1	0	4
R3	9	7	2	7	8	1	6	6	4	7	5	0
R4	0	0	3	3	3	8	7	7	4	9	9	6
R5	0	3	6	5	2	6	0	3	8	5	2	1

You should practice enough and get to the point of figuring out the best entry point. The unconventional table should be used in conjunction with the other tables. The unconventional table is formed based on words and numbers.

This makes it near impossible for the people at the receiving end to figure out your trend. The conventional trend is based on numeric 0123 that could put you at a disadvantage.

You might be tempted to wonder how I could catapult the unconventional table to winning lottery numbers.

Let us examine one section against past Maryland lottery Pick 4 result.

Maryland played 5512 on 9/10/12.

In the unconventional table above I figured that I could create the same winning Pick 4 number by converting the winning numbers in columns F, G, H and I.

If you look at that column, you will see the winning number 8774 in row R4.

The entire group under columns F, G, H, and I when converted will change from,

F	G	H	I
2	4	8	1
9	1	5	4
1	6	6	4
8	7	7	4
6	0	3	8

Into the Shadow Numbers below,

8124

6471

4991

2551

9302

You can see that I have achieved the aim of creating the 5512 that played in Maryland.

The trend is intact because I converted the entire group into Shadow Numbers.

Let me now put our new result down against actual Maryland Pick 4 numbers over the same period to see if the unconventional system works.

8124 Maryland played 8420 on 9/14/12

6471 Maryland played 1467 on 9/28/12

4991 Maryland played 3992 on 9/12/12

2551 Maryland played 5512 on 9/10/12

9302 Maryland played 9203 on 9/11/12

If you take a very close look, you will see the relationships between the dates. You can actually catch some of the winning numbers based on the dates. I have stated earlier in this book that there is indeed a relationship between the dates and actual winning numbers.

The other thing worth noting is the middle winning Pick 4 number 4991. You will notice that Maryland played it as decoy number 3992. The sum of both winning numbers will be 23. The decoy winning number 3992 is meant to alter the entire equation.

Our unique winning methods captured the winning numbers in the trend regardless.

The most powerful thing here is that you can accomplish this with different words and numbers. It is near impossible for anybody or computer to decipher how you worked out the winning numbers in your trend.

The last winning number among the group is often altered by one which is the reason you have the winning number 8420 against 8124 that our unique methods produced. I can also recreate other trends that will capture the 8421 and produce other winning Pick 4 numbers.

You can use this method to gauge the pulse as well as produce winning lottery numbers. You can work out winning trends with the other tables against the unconventional table and actually see the winning numbers.

You can mitigate losing by using more than one table. There is no winning lottery number you cannot work out every single day if you painstakingly comb through the tables and apply the three methods I have been describing throughout this book.

If I tell you without demonstration that you can actually zero in on winning lottery numbers from words, you are likely not going to believe me. The reason of course is because the convention did not teach us that. You are not expected to master numbers to that level.

One thing that is actually more important than winning is studying **Lottery Icon** and mastering the methods. Experiment with the methods. Test them against past results. Test them against different states. Anytime you bet and lose, revisit the trend and try to figure out why you lost.

The winnings will come when you are ready.

I will use the unconventional table against other state lottery result to show you that the table is efficient.

Let me take the method a step higher. I will go to a different state, Texas, and use the exact same date to prove to you once more that this is the most efficient lottery method on the planet.

It is much easier to comb through the entire tables until I see a very comfortable entry point. I am trying to show you that you can create endless opportunities once you master the methods.

Texas lottery played 8819 on 9/15/12.

This will be the same date that I used when I was demonstrating the Maryland lottery trends.

This is important so as to show you that you can create opportunities even if you are left with only one table. I can create endless winning trends with just one table. You can do the same by the way once you master my methods.

Texas lottery played 8819 on 9/15/12.

My goal at this point is to create 8819 from Unconventional Table 2.

I will do this with the winning Pick 4 numbers on columns G, H, I and J.

I chose the columns because they are close to the one I used for Maryland lottery winning Pick 4 numbers demonstrations

I will assume that you know how to do the conversions at this point.

I will convert the winning Pick 4 numbers without elaborate explanations.

I will rather use the time and space to show you different ways of working out the winning trends.

The group in question that I will be converting from Unconventional Table 2 is,

G	H	I	J
4	8	1	0
1	5	4	1
6	6	4	7
7	7	4	9
0	3	8	5

I will display the above group without the grids and place the conversions below them until I achieve the winning Pick 4 number 8819.

Original Group

4810

1541

6647

7749

0385

 The best Pick 4 number that could help me achieve my goal of 8819 is 7749. I will convert the first and second digits into Counterpart, and the third digit into Shadow. I will leave the last digit unchanged. The reason for that is because 7749 has the number 9 as the last digit like that of 8819. I will convert the number 4 into Shadow 1 (one) that would enable me have 19 just like that of 8819.

I will convert the first two digits into Counterpart that would enable me turn the numbers 77 into 22 after which I will apply Shadow and turn the 22 into Shadow 88, thereby creating 8819.

Do not forget to do the same thing to the rest of the group otherwise the trend will be violated.

First Conversion.

9340

6011

1117

2219

5825

Second Conversion (first two digits into Shadow)

6040

9311

4417

8819

7225

I have now completed the conversion and accomplished my goal of creating 8819.

Let us examine the new trend with actual Texas lottery results to show you how to win big money with this methods.

Texas lottery started playing the trend with winning Pick 4 number 4417 on 8/23/12 with the other ones that repeated notwithstanding. You may not at this point know the rest of the trend.

There are other ways but let us concentrate on this trend at this point. The next day, Texas played 6070 on 8/24/12 while our method produced 6040.

What could be wrong with this scenario?

The 6070 is actually being played as 6040.

Please pay very close attention here.

If you understand this part and apply it, you will actually know the potential winning Pick 4 number that is about to play in no distant time.

Remember that a small section of the trend is often played as decoy which would make it difficult for you to work out the rest of the winning numbers.

You have to think critically.

Two days later Texas lottery played 1351 on 8/27/12.

How come our method produced 6040 and 9311 against the two winning Texas lottery Pick 4 numbers that they played (6070 and 1351)?

They indeed played the role of the two we worked out.

Take another look at 6040 and 9311 Vs 6070 and 1351.

The Counterpart of 4 (6040) is 9 (9311).

The Shadow of 7 (6070) is 5 (1351)

You can see that they replaced the two by simple Shadow and Counterpart conversion.

On critical thinking you would have figured this out at least when you saw the 1351.

You can go through the above scenario several times until you get a good grasp of it.

The payout on this would begin immediately because you would have embarked on playing the rest of the trend.

That would be the case because Texas played 7225 straight on 8/28/12. That would be just another big win that would pay out about $5000 for every $1 straight bet.

You can see that 60 percent of the winning Pick 4 numbers in this group played straight.

Pay close attention to the times they alter trends by simple conversion like that of 6040 and 9311.

Look forward to making big money. The reason for that is because you are not expected to understand the game of lottery to that level.

You can attest to that by asking the next one hundred lottery players who have not read this book. I will bet that none of them have the knowledge. The people out there who claim to be writing books on lottery do not know either.

You have acquired more knowledge than all of them combined. I will always stand by you and that statement.

You are witnessing the biggest discovery since the invention of lottery game.

This method does indeed decimate the conventional odd number calculations.

The winning odd ratio is based on the convention of doing the lottery number workouts based on the numeric 0123 and so forth.

The Unconventional Table on the other hand understands that there is relationship between numbers and letters.

Some readers may be tempted to question the veracity of this method.

I will construct one more Unconventional Table to show you the efficacy.

This system is so powerful that you cannot fit it inside the conventional odds calculations.

I have already demonstrated it with Maryland and Texas lottery results.

Let us build the new table together and test it against another state.

Tops

We are going to begin by forming the numbers in the trend in the form of the word, "Tops".

The process 1 will have the numbers in the form of the letter "T".

Process 1

	A	B	C	D	E	F	G	H	I	J	K
R1	0	1	2								
R2		3									
R3		4									
R4		5									
R5		6									

Process 2 will follow by forming the numbers in the form of "O".

Bear in mind that the "O" number formation could be in capital or small letters. It doesn't have to fit exactly like the one I am forming.

They will form winning trends regardless, when you apply the methods in Lottery Icon.

Now let us place the next set of numbers in the grid in form of the letter, "O".

I decided to create it in the form of small letter.

Bear in mind that I have already covered some parts of columns A, B and C in Process 1 above.

I will continue with number 7 since Process 1 stopped at number 6 above. You could start from any number you like, by the way, to make it increasingly impossible for the other side to read your trend.

Process 2

	A	B	C	D	E	F	G	H	I	J	K
R1											
R2											
R3			9	8	7						
R4			0		4						
R5			1	2	3						

You can see that I placed the numbers in the form of letter, "O" above.

The last number that I placed is 4.

I started Process 2 with the number 7 and completed with number 4.

I have now created trend that is formed in the shape of T and O towards forming the entire group based on the word, "TOPS".

I will place the two newly formed group in Process 3 before continuing.

The table will now appear like the one below,

Process 3

	A	B	C	D	E	F	G	H	I	J	K
R1	0	1	2								
R2		3									
R3		4	9	8	7						
R4		5	0		4						
R5		6	1	2	3						

I will continue the process that will form the numbers in the shape of the letter, "P". I will start with number 5 since the Process 2 ended with number 4.

Process 4

	A	B	C	D	E	F	G	H	I	J	K
R1						5	0	1			
R2						6		2			
R3						7	4	3			
R4						8					
R5						9					

I have now formed trend that has the shape of P towards my goal of forming trends with the word, "TOPS".

I will now merge it with Process 3, thereby forming the word TOP with numbers.

The merger will now be Process 5 after which we will be left with creating same in form of, "S" for the completion of our task.

Process 5

	A	B	C	D	E	F	G	H	I	J	K
R1	0	1	2			5	0	1			
R2		3				6		2			
R3		4	9	8	7	7	4	3			
R4		5	0		4	8					
R5		6	1	2	3	9					

We have now formed the word, "TOP" above. You can see that the table is gradually taking shape.

Let us continue by placing numbers in the form of S to enable us create trend that will take the shape of the word, "TOPS".

I will make use of columns I, J and K to form the "S" which will appear like the one in the table below.

The last number I placed in the formation of the "P" shape above is 4. This means that I will continue forming the "S" shape with number 5.

	A	B	C	D	E	F	G	H	I	J	K
R1									7	6	5
R2									8		
R3									9	0	1
R4											2
R5									5	4	3

I will now merge it with Process 5 to create the trend with the shape of the word, "TOPS".

The final merger will look like the table below,

Process 6

	A	B	C	D	E	F	G	H	I	J	K
R1	0	1	2			5	0	1	7	6	5
R2		3				6		2	8		
R3		4	9	8	7	7	4	3	9	0	1
R4		5	0		4	8					2
R5		6	1	2	3	9			5	4	3

Process 6 above represent numbers 0 through 9 in the shape of the word, "TOPS".

Now that you have completed the top task, you will then fill up the rest of the grid with numbers 0 through 9.

I will fill it with letters first to show you how to do it.

You will replace the letters with numbers 0 through 9 afterward. You can do this exercise on a simple sheet of paper to make sure that it is right. Your final product must match the finished one in this book, otherwise go back and redo it.

I will do it with small letters.

The spaces when filled with letters will look like the one below,

	A	B	C	D	E	F	G	H	I	J	K
R1	0	1	2	e	h	5	0	1	7	6	5
R2	a	3	D	g	j	6	k	2	8	o	r
R3	b	4	9	8	7	7	4	3	9	0	1
R4	c	5	0	i	4	8	L	n	q	s	2
R5	f	6	1	2	3	9	m	p	5	4	3

You can place actual numbers beside them. You do not have to do it in a hurry to avoid mistakes.

I will start filling up the rest of the grids with numbers. I will start with number 6. The reason I chose 6 is because the last number that I formed in the shape of "S" finished with 5.

You can start with any number of your choice and follow the same order.

The final product MUST look like the one below,

TOPS 1

	A	B	C	D	E	F	G	H	I	J	K
R1	0	1	2	0	3	5	0	1	7	6	5
R2	6	3	9	2	4	6	6	2	8	0	3
R3	7	4	9	8	7	7	4	3	9	0	1
R4	8	5	0	5	4	8	7	9	2	4	2
R5	1	6	1	2	3	9	8	1	5	4	3

I will now rotate the numbers in the form I explained in the earlier chapters. The reason I chose to do this is to make sure that the numbers that formed the shape of the word TOPS are distributed over all sections of the table.

This in effect form trends in different formats. You can capture consistent winning Pick 4 numbers when you use the table and apply the conversions that I have been discussing through this book.

I will start from numbers in row R1, walk my way rightwards and continue through the rest of the table.

	A	B	C	D	E	F	G	H	I	J	K
R1	2				3				3		x
R2		6		0		5		9		6	
R3			2				0		x		5
R4		1		6		2		1		6	
R5	0				4		x		7		

I have now placed a section of the numbers in the above table up to where you will find Pick 4 number 4662 in columns E, F, G and H of the TOPS 1table. I will continue in the same direction from columns I, J and K (803).

You can see the (803) with the letter X beside them to show you where I continued with filling up the table.

I am not trying to expend too much time in this because I discussed it earlier. I will continue from row R1 right under column K and walk my way in the other direction.

The table when finished will appear like the one below,

TOPS 2

	A	B	C	D	E	F	G	H	I	J	K
R1	2	3	4	8	3	8	7	2	3	3	8x
R2	4	6	1	0	4	5	4	9	2	6	5
R3	9	0	2	5	7	2	0	1	0x	1	5
R4	9	1	0	6	9	2	6	1	8	6	3
R5	0	5	8	7	4	1	3x	9	7	4	7

We have now finished forming table based on the word," TOPS".

Let us put it to test against Virginia lottery winning Pick 4 numbers over the same period (September 2012) that we have been working with.

This unique method is so efficient that it actually produces winning numbers by itself plus the countless winning numbers you will enjoy by employing the three methods that will enable you convert the trends into countless winnings.

Look at our Pick 4 winning numbers from columns B, C, D and E.

B	C	D	E
3	4	8	3
6	1	0	4
0	2	5	7
1	0	6	9
5	8	7	4

From our table above,

3483 Virginia lottery played 3438 on 9/24/12

6104 Virginia lottery played 1046 on 9/21/12

0257 Virginia lottery played 5720 on 9/2/12

1069 Virginia lottery played 1639 on 9/6/12

5874 Virginia lottery played 7245 on 9/24/12

In the case of 1069 and 5874, you can see that Virginia lottery altered the numbers in the second digits of 1069 and 5874 (0 and 8) by changing them into Shadow.

Remember similar example in that of Texas.

They are not necessarily going to play all the groups Verbatim through the month. You should always be prepared to see the altered number and follow suit.

Additional buttress to prove to you that they are in the same trend is the fact that 7245 played on the same date with 3438. It could not possibly be a coincidence.

If you take another close look at the same section you will see another opportunity.

Virginia played 0097 on 9/1/12 followed by 5720 on 9/2/12.

Our table will show you 1069 in row R4 and 0257 above it on row R3 of columns B, C, D and E.

The 1069 is the same as 0097. Remember that I employed similar workout in the Pick 3 section.

The number above 1069 is 0257 that they played on 9/2/12.

This clearly shows you that they will play 6104 above or 5874 below. The 6104 played ten days later on 9/12/12.

You can create trends where none exist and still be able to capture future trends. The reason for that is because this method is outside the confinement of mathematical odds calculations.

Some greedy people will tell you that the Virginia winning Pick 4 numbers spread over entire month. Ask them the last time they won five Pick 4 numbers in one month. They are the people that keep chasing the wind. You cannot have anything worthwhile without some effort.

Nothing in life is free. You can work out winning Pick 4 numbers every single day. Read the last chapter for more information on that.

This book gives you the tools to bet and win beyond the limits of mathematics that is taught in schools.

Let us create trends from the same period and examine this position.

Virginia lottery played 0816 and 5675 on 9/15/12.

Let us create trends that will encompass the two winning Pick 4 numbers, thereby exposing the rest of the hidden trends. You must remember that the groups move in tandem.

I will look for the best section of the tables that will give at least two digits that will be common with the two winning Pick 4 numbers. The primary reason for that is because the two Pick 4 numbers played on the same date.

I can also use one of the winning Pick 4 numbers with one from a different date and expose the winning numbers for that trend.

Let us concentrate on creating trends with 0816 and 5675.

My best entry point after examining the tables will be,

Tops 2 Table, columns H and I as well as columns B and C.

You will see the reason for that in the next few lines.

H	I
2	3
9	2
1	0
1	8
9	7

In the above columns H and I you can see that if I apply Shadow to the groups I will create 80 from the numbers 2 and 3. I noticed that I will equally create 6 and 5 from the numbers 9 and 7.

This would accomplish one half of the task because I have the two common digits in 0816 and 5675 that Virginia played on 9/15/12.

Let us convert the entire group in the above table.

The result after Shadow application will go from the table above to the one below,

H	I
8	0
6	8
4	3
4	2
6	5

You can clearly see 8 and 0 (zero) as well as 6 and 5 in the above table.

We are now left with creating the remaining digits which would be 1 and 6 to go with the 8 and 0 already created as well as 7 and 5 to go with 5 and 6 that we have already created in the above table.

I chose the table below because I can accomplish that task by converting the entire group into Counterpart.

Let us convert the entire group into a new table that will have the numbers we are looking for.

B	C
3	4
6	1
0	2
1	0
5	8

The above table will change to the new one below after the application of Counterpart.

B	C
8	9
1	6
5	7
6	5
0	3

You can see that we have accomplished the goal of having 1 and 6 as well as 5 and 7 from the same table.

Let us write down the newly created numbers and merge the two with the goal of creating 0816 and 5675 thereby exposing the rest of the trend.

H	I	B	C
8	0	8	9
6	8	1	6
4	3	5	7
4	2	6	5
6	5	0	3

I will bring the two groups out from the above grids before realigning them.

HI	BC
80	89
68	16
43	57
42	65
65	03

I will now realign the group so that I can create the 0816 and 5675.

You will notice that 80 started the H and I columns and ended with 65.

The B and C columns on the other hand started with 89 and ended with 03.

In order to properly merge the two groups, I will start the B and C columns with number 1 and 6 that will be placed beside the 8 and 0. I will count in the other direction so that 5 and 7 will become the last thereby merging with 65.

Let me put the two groups right below each other and realign them. This might be easier for some readers to comprehend. I will put the B and C group below the H and I to effect the necessary change.

HI Group 80, 68, 43, 42 and 65.

BC Group 89, 16, 57, 65 and 03

I will keep the HI Group constant and shift the BC group leftwards from 16 to create the 0816 and 5675 respectively.

If you shift the BC Group from 16 the new BC Group will then become,

16, 89, 03, 65 and 57.

Let me bring the HI Group down along with the newly realigned BC Group.

HI Group 80, 68, 43, 42 and 65.

BC Group 16, 89, 03, 65 and 57.

You can see that 16 is right below 80 and 65 is right above 57.

The group when finished will appear as the trend below,

8016

6889

4303

4265

6557

We have now accomplished the task of creating a trend with the winning Virginia Pick 4 numbers 0816 and 5675 thereby exposing the rest of the group in that trend.

The next task will be to see if the rest of the trend played to confirm that the trend is in fact intact.

I will place the dates that the trend played below.

0816 Virginia played on 9/15/12

8689 Virginia played on 10/28/12

3403 Virginia played on 11/29/12

2465 Virginia played on 12/1/12

5657 Virginia played on 9/15/12

The 0816 and 5675 is of course given since we started creating the trend with those two winning Pick 4 numbers.

They both played on 9/15/12 regardless.

You will notice that the trend played the winning Pick 4 numbers spread out over a short period of time.

The winning Pick 4 numbers played in September, October, November and December.

One other important thing worth noting at this point is that 2465 played in just two days after 3403.

Take another careful look at the dates and see if you will notice anything significant.

I am sure you can at this point, otherwise endeavor to study this book more.

In the above instance I created winning trends from just one day result.

You can do that with one table when you get very good at it or create trends from different tables.

You do not have to be tempted to go the easy route.

There are times the groups may not align properly.

You will notice that most of the tables are in odd numbers.

Let us say for instance that your trend is 1, 2, 3, 4 and 5 but do not fit in the trend you are creating.

The same numbers could be changed into 1, 3, 5, 2 and 4.

You can see that I still maintained a working trend. In the later case you will be able to capture the rest of the winning numbers with 1 and 3 trend that would not have been possible in the initial group. That is precisely why the columns are in odd numbers.

89

16

57

65

03

The group of numbers above could be changed when necessary into the ones below,

89

57

03

16

65

You cannot do that with 6 table columns.

You cannot change the group 1, 2, 3, 4, 5 and 6 into any other trend.

TOPS 3

	A	B	C	D	E	F	G	H	I	J	K	L
R1	7	6	5	7	7	2	5	4	3	1	2	0
R2	6	4	0	6	4	9	1	3	6	3	8	3
R3	3	1	8	0	0	8	1	7	5	9	9	2
R4	2	0	5	2	7	8	8	3	4	7	6	9
R5	1	4	9	8	6	1	2	5	5	0	4	9

TOPS 4

	A	B	C	D	E	F	G	H	I	J	K	L
R1	8	7	1	2	7	7	3	7	0	9	6	5
R2	1	3	9	7	2	2	4	3	4	4	2	0
R3	6	9	5	5	9	3	5	1	0	1	2	4
R4	5	6	0	8	8	4	9	4	6	1	0	8
R5	7	2	3	8	0	6	6	8	3	5	1	9

Pick 4 Galore

Take a very hard and patient look at the Pick 4 galore against the trend that your market is playing to make sure that you extract all the Wealth.

7341	4240	3393	0097	4815
7681	0154	2465	3078	1546
6214	7986	6608	6325	6642
4981	6314	8682	5373	7744
0144	7681	0469	2389	3800
9098	5751	6761	6812	8539
1463	6848	7440	0764	6251
6360	8978	1607	3768	7278
4751	4010	3153	8441	0333
0098	0615	9308	5970	3652
8663	6098	9644	2951	5996
8544	3098	8386	1602	0885
2501	6717	7431	0094	1274
3064	3506	8880	1604	1253

Chapter 4

Pick 5 Lottery Book

The pick 5 lottery is played by some states including Washington DC, Pennsylvania, Georgia and some other states under different names.

Pennsylvania pick 5 is called Quinto, Georgia pick 5 game is called Georgia Five and that of Washington DC is called DC 5.

The pick 5, as the name implies, comprise of five numbers such as 12345 or 89215, whatever the case may be.

The odds of winning the jackpot are 1:100000.

The top prize could range from about $10000 for that of Georgia Five to $50000 for DC and Pennsylvania markets.

The odds, of course, are based on betting on the game based on the numeric numbers 0123 and so forth.

I will give you actual examples and show you how to use them to your advantage against the conventional odds calculations.

Your chance of winning big in this high paying lottery game is superb when you apply the three methods that I have been discussing with the 100% efficient tables.

The tables with the application of Numbers, Shadow and Counterpart actually expose the potential winnings.

One other important thing that is overlooked by so many players is the fact that there are ample ways of winning on pick 5 lottery.

In Georgia Five, you can actually win with just one number out of the five. You can win some money if you catch the first or the last number. That will, at least, recoup your cost of betting on the game and some extra change to come back for the big money.

In DC and other markets you will equally begin to enjoy some winnings by catching just two of the front or back pairs.

You should always consider rotating the numbers and betting from the low odds like front and back pair to the entire five numbers for the jackpot.

If you win front or back pair in DC, for instance, on $1 bet, you will receive about $50. This means that your $1 now puts you in the driver's seat for the next fifty games. You have fifty more chances of catching the jackpot.

Let us assume, for instance, that you are betting on the numbers 12345. You should consider betting 12 in front and back pair as well as 45. You can equally move the middle number 3 to the front and back digit position and employ the parlay method.

Your chance of winning substantial money could not be better because you have the 100% efficient tables and the three methods in your favor.

The 100% efficient tables do not fit in the mathematical odds calculations of any game. The trend is difficult, if not impossible, for anybody or computer to decipher. That is what gives you an enormous advantage.

Let us now do actual example with the TOPS Tables.

You must develop the ability to think outside the box when it comes to betting on lottery games with high odds and huge potential winnings. You are expected to follow the crowd and kowtow to the convention.

How do you create consistent winning pick 5 trends?

There are several ways.

Let me start this with DC 5 from the same date of 9/15/12.

I chose this date to prove to you that you can actually create consistent winning pick 5 numbers every single day. It will be very easy for me to scan through the entire tables to the one that matches one convenient date. Some readers may have that luxury but I don't because I have to show you different ways of walking into the winning numbers from the same date range.

DC 5 played 86795 and 40305 on 9/15/12.

Let me begin the process by creating a more difficult entry point. Please bear in mind that there are other easier ways of creating the trends.

I am trying to recreate 86795 and 40305.

I can use similar numbers from the tables or create similar numbers with Counterpart and Shadow. You are expected to begin the process by taking numbers that will match those from the tables.

The Shadow and Counterpart on the other hand means that nobody really knows where you may chose to begin the process of creating the trend.

The numbers 31 could represent the first two digits of 86795 when Counterpart is applied and the numbers 29 could equally represent 86 when Shadow is applied.

The same thing applies to the rest of the pick 5 numbers.

You should make effort to find where minimum of two common numbers in the trend appears or create one.

In the case of 86795 you should endeavor to find where 86 or more are together. In this case the two common numbers from 40305 should be below the 86 or any other trend you are forming.

This process led me to TOPS Table 3, columns I and J below,

Tops 3

I	J
3	1
6	3
5	9
4	7
5	0

I chose it because you will find the numbers 31 and 59 among the groups. I can change 31 into 86 with Counterpart application as well as change 59 into 04 with Counterpart application. The only slight headache at this point is that 59 is not right below 31.

You can cure that headache by realigning the groups so that 59 will be right below 31 in the same manner that I discussed earlier.

You do this by putting 31 down, skip the next group (63) and put 59 down, skip the next (47) put 50 down, skip 31 (that is already done) and put the next (63) down and finally 50.

Let me bring the groups out from the above table and put down the rearranged group below that to accomplish my goal of having 59 right below 31.

The above group without the grids,

31

63

59

47

50

The same group after rearrangement that will have 59 right below 31:

31

59

50

63

47.

I have now completed the task of getting some (40%) of the required digits I will need towards creating 86795 and 40305 trend.

I will leave the new group at this point and continue looking for where to get the rest of the group.

The search led me to TOPS 1 Table, columns I, J and K.

TOPS 1

I	J	K
7	6	5
8	0	3
9	0	1
2	4	2
5	4	3

If I convert the above groups into Shadow the 765 will change to 597 to compliment the 86 while the 803 on the other hand will produce 230. The 230 will create problem in this case because I need 5 instead of 2. Remember that I am trying to create 305 to compliment the 40 from 59 thereby making it 40305.

In this scenario it means that I will need to find or create 7 in place of 8 that would produce 305 as against 230.

In other wards I will need to find or create trend that will have 7 and 7 under the I(i) column. I will go back to the tables and see where to find it or create one.

I can accomplish that by converting the group under TOPS 4 column K as well as TOPS 3 column F.

The numbers under TOPS 4, column K are,

6

2

2

0

1

The same numbers when converted into Counterpart will become,

1

7

7

5

6

As you can see I now have number 7 right below another 7.

Let us complete this part before converting the second set.

I will bring down the table and replace the I column with the newly formed group.

TOPS 1

I	J	K
7	6	5
8	0	3
9	0	1
2	4	2
5	4	3

The same will now appear as the one below when the I (i) column is replaced with the newly formed group.

TOPS 1

I	J	K
7	6	5
7	0	3
5	0	1
6	4	2
1	4	3

You will notice that I started the replacement with number 7 to align with the J and K column groups properly.

I will now bring them out from the table and merge them with the first two digits groups that I formed earlier.

31765

59703

50501

63642

47143

The first two digits will be converted into Counterparts and the last three will be converted into Shadow. I will convert the first two digits from left to right and convert the last three from right to the middle digits.

If the explanation is getting a bit heavy, pause for two minutes before you continue. I intentionally chose the difficult route to prepare you for different scenarios. There is nothing worth having without good effort.

The first two digits will now appear as follows after Counterpart application,

86

04

05

18

92

The last three digits after Shadow conversion from the other direction will appear as the group below,

795

035

437

819

014

I will now merge the two groups into pick 5 and the result will be like the one below,

86795

04035

05437

18819

92014

I have now completed the trend with DC 5 winning numbers 86795 and 40305.

I will repeat the process with the numbers from TOPS 3, column F. The process is more or less the same except for the fact that I am converting the group from column F that will change the numbers 8 and 8 into 7 and 7.

I am sure you can do this by now. I will write the group down, convert them and replace the ones we just finished forming with the new one.

The primary reason for this is because you could be confronted with more than one opportunity when you are creating trend and deciding on which one to choose.

It will become apparent as you read further.

I will just put the numbers from TOPS 3 column F down.

I will change the numbers into Shadow and then Counterpart.

I am going to write them beside each other so as to save time.

2	8	3
9	6	1
8	2	7
8	2	7
1	4	9

The earlier group that we converted was,

31765

59703

50501

63642

47143

I will now replace the middle digits with the new group that has 7and 7 after which the above group will appear as follows,

31765

59703

50901

63342

47143

The new group when converted in the same manner we did the other one will become the group below,

86795

04035

05436

18810

92014

I have now completed the conversions of both groups.

Let me put down side by side the finished trends before proceeding further.

86795	86795
04035	04035
05437	05436
18819	18810
92014	92014

You now have the two finished groups side by side in the above table. The trend shows you that the next potential winning DC 5 number is 05437 or 05436 if they continue in the same direction or 92014 if they chose the other direction.

The way to play those will be by employing the front and back pair as well as the other options. You can win pick 5 with two or more numbers.

The lower winnings keep you in the game without expending much from your pocket through your journey to the $50000 jackpot.

Something worth noting is that this trend is developed outside the confines of the 1:100000 odds convention.

In the popular convention you need to spend about $100000 to win $50000. That of course is based on the assumption that you did not make any mistake.

You can go for the same goal by betting one number on same pick 5 numbers game twice daily. If you choose the latter route on $1 bet, it will take you about 137 years if you did not make any mistake. That would be a long time because so many lottery players would like to enjoy the winnings at a much younger age anyway.

I don't know if you will call that winning anything because inflation would have adjusted that winning to zero coupled with the fact that it is a long time.

You can choose the unique methods in **Lottery Icon** and go for the big money over a short period of time.

The way to know if it works is of course by verifying the newly created trends.

The 86795 and 40305 already played in DC 5 on 9/15/12.

Go through the results after the above two winning DC 5 numbers. You will notice that the remaining pick 5 numbers in the trend would have given you some money. You will see that three or more digits among the groups played several times over the next three weeks including the 18851 that played on 9/25/12.

That will of course be ten days after the original winning numbers that started the trend. The trend paid money all the way till the entire five numbers played.

DC 5 played 54036 on 10/5/12.

You can see that this unique method produced the five required winning numbers to win the $50000 jackpot in less than three weeks.

You should also make adjustments as necessary.

When they played 18851, you should recreate that column or find one from the tables that could recreate the newly introduced number.

My trend had 18819 instead of 5. You will of course capture some winnings and proceed to create trends with 5 instead of 9. The 5 when introduced will expose new winning pick 5 numbers.

Let us recreate and look at some trend against DC 5 winning numbers. In the above instance we started with winning DC 5 numbers from 9/15/12 so as to be in the same time frame with the other examples in this book.

We will now look or recreate trends with the results from the next day 9/16/12 in mind. I do not want to saturate the same date of 9/15/12.

We will do this with TOPS Table 4.

TOPS 4

	A	B	C	D	E	F	G	H	I	J	K	L
R1	8	7	1	2	7	7	3	7	0	9	6	5
R2	1	3	9	7	2	2	4	3	4	4	2	0
R3	6	9	5	5	9	3	5	1	0	1	2	4
R4	5	6	0	8	8	4	9	4	6	1	0	8
R5	7	2	3	8	0	6	6	8	3	5	1	9

At this point I would like to recreate the 70236 that played on 9/16/12.

I chose columns F, G, H, I and J.

I will cut and paste that section before embarking on the conversion.

TOPS 4

F	G	H	I	J
7	3	7	0	9
2	4	3	4	4
3	5	1	0	1
4	9	4	6	1
6	6	8	3	5

In the above table, you can see that I can recreate 70236 from the top pick 5 number 73709. I will not do elaborate explanation on the conversion at this point.

The numbers from column F will come down as they are because the first digit 7 satisfies the need for the first digit of the DC 5 number (70236) that I am trying to recreate. The second digits will be converted into Shadow because it will satisfy the second digit (0) of the 70236. The third digits (7) will be converted into Counterpart from 7 to get the required number 2. The fourth and the fifth digits columns (I and H) will be converted into Shadow to get the required numbers 3 and 6 from 0 and 9.

The above table will now change after the conversion to the one below,

70236

21811

37634

46994

69307

We have now completed the process of recreating the trend with the winning DC 5 number 70236 that played on 9/16/12.

I will put the group down alongside the ones that played in DC 5 with actual dates. That will be another buttress on the argument that the TOPS Tables are 100% efficient.

70236 DC 5 played 70236 on 9/16/12.

21811

37634 DC 5 played 33764 on 10/2/12.

46994

69307 DC 5 played 67039 on 11/11/12.

In the trend we created from our table DC 5 played three pick 5 numbers in its' entirety. Each of those would pay you $50000 if you get the five numbers straight. You equally get paid handsomely if you get the numbers in ANY order. You will of course get paid a lot more if you follow the methods that I described earlier in this chapter.

Another important point that must not be omitted here is about the other two DC 5 numbers in the trend.

If you go back to DC 5 results, you will notice that DC 5 played 81221 on 9/29/12 while our trend produced 21811. The 21811 would equally give you winnings except the jackpot.

Another DC 5 winning number 49064 that was similar to the trend our system produced (46994) played on 11/15/12.

On the two pick 5 numbers in question, 21811 compared to 81221 that DC 5 played and 46994 compared to 49064 that DC 5 played.

In both cases you will see that if you add one (1) to one of the digits in the ones our table produce, you will get the same pick 5 winning numbers.

Suffice this to mean that you could use any pick 5 numbers from our tables, add or subtract one from any of the digits and you will be on your way to exposing a lot of winning pick 5 numbers that are about to play.

The trend that we created by itself give you additional opportunities. Take a look at the 81221 that played on 9/29/12. The next day DC 5 played 76384 on 9/30/12.

Our trend on the other hand produced 21811 and 37634. You can see the similarities between the two groups.

You could make good money on the 76384 from having the trend from our table in your hand.

If your budget permits, you can take any pick 5 number from the TOPS Table, add and minus one (1) from each of the numbers from the trend you are creating and in effect expose ten different trends

An example will be creating the groups under 70236 as the original trend and changing the 70236 into the following numbers by addition and subtraction of one (1) on each of the digits, (80236, 60236, 71236, 79236, 70336, 70136, 70246, 70226, 70237 and 70235)

Each of the above will play the same role as 70236. You will see different trends under the new groups. You can watch the groups against actual results and make the adjustment and changes when necessary. They will expose the winning pick 5 numbers consistently like that of 81221 and 76384.

Chapter 5

Lotto Winning Secrets Book

The naysayers will tell you that the lotto could not be won. A part of their reasoning could be based on the fact that you have lotto games with odds calculations of upwards of 259 million to win. Those odds are, of course, based on the popular convention. I will not advise you to belabor on convincing them otherwise.

I will put the energy in showing you that you can indeed win the lotto games.

Pay close attention to learn how to break the odds from this point on. This section is one of the most important parts of **Lottery Icon**.

Before I delve into it, let me touch on a small section from one of my lotto books that will help you.

The book in question is called,

Powerball, Mega Millions, Euro Millions, LottoMax Formula.

If you go to page 53 of the above mentioned book, you will see the Precision Board 9.

The boards did not earn that name by mistake. The calculations are done to give you precise winning numbers.

You will see the following winning Powerball numbers from Precision Board 9,

9, 23, and 40.

The above lotto numbers played recently on 7/27/13.

The next sets of lotto winning numbers beside those are,

9, 19, and 32 that played on 8/28/13.

Take another close look at the dates the two winning groups played.

The first played on 7/27/13 and the next played on 8/28/13.

This feat could only be achieved through precision methods.

You should take a look at the next sets of numbers and begin to make profitable use of them.

The Precision Boards cover all the numbers.

I will not spend too much time on the above book. I do, however, encourage you to study the book once more. The Precision Boards will become extremely handy.

The reason will become apparent at the end of this book.

Let me, at this point, touch on some past lotto winning numbers and show you how to read and make good use of them.

I will go back to the same period of time that I have been using throughout this book.

There is a very prevalent relationship between past and present lotto results.

If you look at the Powerball results from 9/1/12 you will see winning number 8 among the groups. The next Powerball result produced winning number 4 on 9/5/12 followed by winning Powerball number 6 on 9/8/12.

The above winning Powerball numbers played in the order of 8, 4 and 6.

You have to master the art of thinking ahead with respect to what these numbers are going to play.

In the first two results you would have winning numbers 8 and 4. The natural thing would be anticipation of winning numbers 2 or 16.

The reality is that you should carefully study all of the potential winning numbers.

In this case the third result produced winning number 6.

If you rearrange the groups, the winning numbers will become 4, 6 and 8. You can see that the trend will begin to make more sense. The numbers 8 and 4 are meant to throw you off guard.

In this case you should be able to capture the winning number 6 on 9/8/12.

I will revisit the winning Powerball numbers you are expected to capture and show you how to make use of them later.

The following numbers also played on the same dates,

21 on 9/1/12 followed by 19 on 9/5/12 and finally winning Powerball number 20 on 9/8/12.

Once again you will notice that the numbers played as 21, 19 and 20.

If you rearrange the group properly the numbers will be as follows,

19, 20 and 21.

You are expected to capture the third winning Powerball number (20) in the trend.

There are trends that stretch longer, too. They are not going to make it easy for you to figure them out.

You will find one of such with winning Powerball number 29 on 8/22/12 followed by 49 on 8/25/12, another 49 on 8/29/12 and the third 49 on 9/1/12. You can see that 29 started the trend and has, so far, played only once.

Let me give you the benefit that you may not know at this point; 29 is part of the trend until 9/5/12 when it played for the second time. It will make sense that you should consider playing it for the third time to compliment the winning Powerball number 49 that played three times so far.

This means that you would capture the winning number 29 on 9/8/12.

You must also look at the winning trends from different angles and take advantage of the opportunities when they present themselves like that of Powerball winning number 48.

The winning Powerball number 48 played on 9/8/12 and repeated in the next result on 9/12/12. You should at this point consider playing 48 on the third result of 9/15/12. The reason for that is because 48 is the next number to 49 that just played three times right after each Powerball result.

Guess what, 48 played on 9/15/12.

You should also employ the Power of Shadow and Counterparts. Remember that I told you throughout **Lottery Icon** that no trend can escape your grip once you master how to use it. You will be able to see what others cannot.

An instance was the Powerball winning number 3 that played on 9/15/12 followed by the Counterpart winning number 8 on 9/19/12. You will, of course, know that the next number should be the Shadow of 8 which is winning Powerball number 2 that played on 9/22/12.

You will also notice on the same dates Powerball winning numbers 43 on 9/15/12 followed by 23 on 9/19/12 and 33 on 9/22/12. You are expected to capture the winning Powerball number 33.

If you rearrange them properly the numbers will go from 43, 23 and 33 to numbers 23, 33 and 43.

You will notice that the winning Powerball number 48 served dual purposes at this point. You will see that in various places like winning Powerball number 3 on 9/15/12 followed by 1(one) on 9/19/12 and 2 on 9/22/12. The numbers when properly arranged will change from 3, 1 and 2 into 1, 2 and 3.

You will also watch out for and capture longer trends when they appear. One of such will be the following winning Powerball numbers from 9/15/12 through 9/26/12.

You will see the following winning numbers 3, 23, 33 and 13.

The numbers when properly arranged will look like the following,

3, 13, 23 and 33. You are of course expected to capture one or more of the numbers in this trend.

Let me raise the bar a little bit on how to capture more winning lotto numbers from the knowledge you acquired throughout **Lottery Icon**. This knowledge separates you from the crowd.

You will find winning Powerball number 26 that played on 9/5/12.

I will do a complete conversion on the winning number 26 and explain how to use it.

The conversion will go from Number to Shadow to Counterpart.

26 -89 -34 -01- 56- 79 -24 -81 -36 -09 -54 -71 -26

You will always arrive at the number you started with at the end, otherwise go back and redo it.

The complete conversions derived from winning Powerball number 26 that played on 9/5/12 are,

26 -89 -34 -01- 56- 79 -24 -81 -36 -09 -54 -71 -26

Now let us take a close look at the conversion and see if we can capture some winning number (s) from the group.

26 played on 9/5/12. They skipped 89 and played 34 in the next Powerball result of 9/8/12. They also skipped the next number 01 (one) and played 56 on 9/12/12. They continued the trend by skipping number 79 and playing winning Powerball number 24 as well as skipping number 81 and playing winning Powerball number 36 on 9/12/12.

The last three fall under what I call an exhaustion point.

You can see that you had adequate time to see the trend and capture the last three winning Powerball numbers 56, 24 and 36 on 9/12/12.

The readers who followed my work before will be able to catch the last three winning numbers easily.

This method is also powerful enough to expose the winning lotto trends before the actual results.

One of such will be doing a conversion with winning Powerball number 21 of 9/1/12 or any other number for that matter.

Let us convert the winning Powerball number 21 and see what happens.

The conversion will also be like the one we did earlier.

21 -84 -39 -06 -51 -74 -29 -86 -31 -04 -59 -76 -21

Make every effort to develop the ability to see the trend. There is a reason why each and every single lotto number played.

You may recollect that I mentioned 29 earlier when it complimented the winning Powerball number 49 that played three times.

In the above conversions you will notice that number 29 played with 31 on 8/22/12. You are not expected to remember that at this point. You may not notice that the two are related unless you do the conversions.

Pay close attention.

I am going to touch the same Powerball numbers that played on 8/22/12.

There is another winning number you might have missed on your interpretation of the trend.

The Powerball winning numbers that played on 8/22/12 actually included 74 from the above conversion that was played backwards as 47.

You could safely say that the conversion produced the following winning Powerball numbers: 29, 31 and 47 on 8/22/12.

This means that three winning Powerball numbers played from our conversion on 8/22/12. You should pay very close attention to numbers like 74 that were played backwards. Some readers may think that the 47 did not come from 74.

Keep on reading to see why you will become the best if you carefully study **Lottery Icon**.

The same trend continued by playing 4, 29 and 51 on 9/5/12 followed by 29 and 6 on 9/8/12 (plus 84/48).

You will notice also that the Powerball results of 9/5/12 produced three winning numbers like that of 8/22/12.

The common number among the two results at this point is winning Powerball number 29. Your job is to follow the same trend and look at producing two or more winning Powerball numbers with common number 29 being one of them.

In the above line you can see that Powerball result of 9/8/12 produced numbers 6 and 29. Your job is not done yet. The prior results are always part of the mirror you need to capture the next winning numbers.

If you recall what I said about 74 playing as 47, you will notice that Powerball number 48 also played on 9/8/12. That of course will give you the three winning Powerball numbers like the other earlier results. The winning Powerball number 48 did not come out of thin air.

You will also notice number 84 from the conversion that was played as 48. The winning Powerball number 47 (74) is the clue that would have given you the winning number 48 (84).

In this instance you could continue with the trend.

You can convert another number that will produce winning number 49 (94) as well as 46 and other winning numbers in that trend. The reason I mentioned 46 is because they played 47 followed by 48. They could continue in the same direction that would produce 49 or go the other way that would be 46 before the winning number 47.

Let us bring the conversion down and take another close look on the winning Powerball numbers.

21 -84 -39 -06 -51 -74 -29 -86 -31 -04 -59 -76 -21

29, 31 and 47 played on 8/22/12.

47 and 29 are together. They skipped one number (86) and played the next 31.

4, 29 and 51 played on 9/5/12.

They continued in the same direction (rightwards) by picking the winning number 4 next to 31 that already played. They did the same thing leftwards by picking the winning number 51 that is next to 47 (74). The winning Powerball number 29 is constant.

They completed the trend by picking 29 and 6 in the same direction. The two numbers are given.

6, 29 and 48 played on 9/8/12.

In the third result you are expected to capture some of the winning Powerball numbers. You are likely going to pick winning numbers 59 and 6 based on prior results. It may not necessarily work out that way. That is the singular reason why you should master how to look at the trend from different ways.

Bear in mind that the winning trend on the third group could be slightly different. You could have gotten numbers 6 and 29 based on prior results. You should also capture number 48 if you had carefully studied the 47 (74) scenario. That should be one reason to consider winning number 48 (84) from the conversion that would have netted three winning Powerball numbers on 9/8/12.

You could easily capture more winning lotto numbers by having more conversions and studying them I might add.

The last three winning Powerball numbers came from just one conversion.

You can see how I am gradually catching all the winning numbers based on different solid winning methods. There is nobody that can challenge you when it comes to numbers if you properly study this book. I know enough to answer any questions when it comes to lottery.

You will reap the reward if you do the work.

You know at this point that you could indeed capture some of the winning lotto numbers when you apply the methods I have discussed so far.

You can use the other lotto book I mentioned earlier to cover any gaps. The precision boards from the other lotto book will give you endless advantage.

You will now be aiming to shatter the entire lotto odds.

The key to that is by making use of the Tops Tables.

I stated earlier that I would revisit the Powerball winning numbers that you are expected to capture.

Let us take the identified winning Powerball numbers from the results of 9/8/12.

We clearly identified the following winning Powerball numbers,

6, 20, 29 and 48 among the winning Powerball numbers that played on 9/8/12.

Based on the results the only remaining winning Powerball numbers to complete the group would be 34 and 44 to capture the entire winning Powerball numbers for 9/8/12.

I could give you a breakdown of each and every single winning Powerball or any other lotto numbers. That task would take entire new book and it is beyond the scope of this book. I will consider that project if enough readers would like to become lottery number maestro.

More details later.

You could reach me through any of the social media for more details. I will also be making announcements as necessary through the website Lottery Sun.

At this juncture we have identified the Powerball winning numbers,

6, 20, 29 and 48.

We are now hunting for the other two required numbers to break the Powerball 175 million odd calculations and go for the Jackpot. I must also state that you do not need to catch all six numbers to win millions of dollars. You can easily do that with five winning numbers. The lotto multiplier is always an option you could take advantage of.

We will now employ the Power of Tops Tables to capture the remaining two winning Powerball numbers (34 and 44).

Let us bring them down and resume the task of identifying the two winning Powerball numbers.

Tops Table 3

	A	B	C	D	E	F	G	H	I	J	K	L
R1	7	6	5	7	7	2	5	4	3	1	2	0
R2	6	4	0	6	4	9	1	3	6	3	8	3
R3	3	1	8	0	0	8	1	7	5	9	9	2
R4	2	0	5	2	7	8	8	3	4	7	6	9
R5	1	4	9	8	6	1	2	5	5	0	4	9

Tops Tables does two things that many would consider mathematically impossible. It actually merges the winning lotto numbers with the trends that could be derived from lottery number conversions (Numbers, Shadow and Counterpart).

No lottery trend can escape the iron grip of the number conversions. In the event that you can't

Identify the rest of the winning lotto numbers; the Tops Tables will come handy. The two when used properly destroys the conventional odd calculations of 175 million plus in the case of Powerball and does the same thing with every other lotto game.

You can trace the potential winning Powerball numbers with prior results and you can also do it with the identified winning Powerball numbers (6, 20, 29 and 48).

Once you capture the winning Powerball numbers you'll be on your way to the big money. There are two options here. If you are working on tight budget your options are few. You will need to make use of the methods I have so far described meticulously.

If on the other hand you are flexible budget wise you could shake the entire lotto system.

The schools walls teach everybody nearly the same thing. You are expected to operate within the confines of those walls. That convention teaches everybody to look at numbers from the angle of numeric order like 123. The lotto on the other hand is not necessarily played based on 123 order. I have shown you examples of where Powerball played the numbers 312 that when arranged in numeric order would be 123.The Powerball jackpot winning odds of over 175 million and other lotto games as well are based on the convention that you would operate based on 123 order that you studied in the school walls.

If you step outside those walls you will notice that the Tops Tables defies those odds logic. Take another look at columns I, J and K of the above table. You will see in the first row winning Powerball numbers 3, 1 and 2. The columns clearly shows those numbers in the same format that the Powerball numbers are drawn.

It does not mean that they are not going to play 1, 2 and 3 numeric order at some point. The table is capable of recalculating and capturing the numbers in that trend when that happens.

In the above case the Tops Table produced the winning numbers 3, 1 and 2 in columns I, J and K. Some readers may wonder about the relevance. Take another look at the above table. The numbers that played below the 3, 1 and 2 are numbers 6, 3 and 8.

The numbers do indeed give you clue to the next winning lotto numbers. The above winning Powerball numbers played through September 2012 that we have been discussing. The trend is so efficient that it produced the above winning Powerball numbers in order. That of course is possible because the format zeroed in on the order the lotto games use.

The number 8 played on 9/1/12.

The number 6 played on 9/8/12.

The number 3 played on 9/15/12.

The number 8 played on 9/19/12.

You can see the numbers 6, 3 and 8 that started the trend. In the first three results they skipped one Powerball result. Go to your Powerball results and check. At the third result the numbers 6, 3 and 8 or (863) as played indicates that the numbers 3, 1 and 2 is about to play based on the Tops Tables.

They played them right after each other.

This means that the winning lotto numbers are closely connected when you use the Tops Tables.

Take another hard look at the above table from columns E, F, G and H. We are looking at rows R1 and R2 in each case. I will write down the numbers beside the letters in each column.

E (74 or 47), F (29 or 92), G (51 or 15) and H (43 or 34).

Another close look will show you that the above numbers matches the ones I discussed earlier that we derived from the conversions.

The 47 played with 29 on 8/22/12.

The 51 played with 29 on 9/5/12.

The 34 played with 29 on 9/8/12.

You can see that you would have captured winning Powerball number 34 on 9/8/12 thereby giving you five winning numbers. We discussed earlier the captured winning Powerball numbers 6, 20, 29 and 48. You will now include 34 in the Big Money Bag.

If you are on tight budget you may not be able to capture the sixth winning Powerball number (44) from the above table. Five numbers will pay you handsomely and give you room for flexibility in upcoming lotto games.

If on the other hand you are flexible budget wise, you could easily capture winning Powerball missing number 44 from the table below.

The reason for that is that you can catch all six numbers mathematically. You have a total of five numbers at hand. If you pick the numbers around the captured numbers or numbers close to last Powerball result you will find the winning number 44 among them.

You need a total of 18 numbers that would produce 42 games that could give you several winning Powerball numbers. You could also capture the jackpot by using the Tops Table below.

I will employ the Power of Tops Tables based on the assumption that you have some flexibility. You can mathematically capture all winning six numbers that would give you the jackpot when you apply the method I am going to discuss from this point.

Tops Table 4

	A	B	C	D	E	F	G	H	I	J	K	L
R1	8	7	1	2	7	7	3	7	0	9	6	5
R2	1	3	9	7	2	2	4	3	4	4	2	0
R3	6	9	5	5	9	3	5	1	0	1	2	4
R4	5	6	0	8	8	4	9	4	6	1	0	8
R5	7	2	3	8	0	6	6	8	3	5	1	9

The goal remains to pick about 18 numbers in total and wheel the numbers to capture most or all of the required six numbers. You can use free lottery wheel systems that is all over the internet.

We have thus far captured four numbers,

6, 20, 29 and 48.

If you select the winning Powerball numbers from the above two Tops Tables, you could equally include the winning number 34 thereby making the captured numbers five (5) in total.

Always remember that you can pick the numbers from more than one Tops Table.

Let us at this juncture use the above Tops Table 4 only.

As I stated earlier one of the options is to pick the winning Powerball numbers from the table that are close to the ones you have already captured and the other option is to pick the numbers based on prior lotto results.

Let us use the former at this point.

From the above Tops Table one of the winning Powerball numbers we captured is winning number 20.

You will find winning number 20 in row R2, column K and L.

You will also see the winning Powerball numbers 34 and 44 that could have been among your total of 18 selections that would have given you all the six numbers for 9/8/12.

In this case you might have included numbers 3, 4, 43, 34, 44, 24 and number 48 that is below the number 20. The 48 may not be necessary since you captured it already. It is of course good to know that this method would have caught it if you did not capture it earlier. You also would have captured number 6 if you work with number 48/84 that is in row R4, columns E and F as well as 29 that is above it.

Tops Table is so Powerball that when you have some flexibility, it catches all the required winning lotto numbers at ease. We have only selected about half of the 18 Powerball numbers and still caught all the numbers.

It is almost impossible to lose when you look at the lotto prize breakdown.

Take a look at the Powerball Prize and Odds Table against the numbers we captured and the results of Powerball on 9/8/12.

You will see that you are bordering in the multimillion dollar territory.

The five winning Powerball numbers would pay a cool million dollars.

If you captured the Powerball number (29) you will be enjoying the Big Jackpot. I am sure that you will testify that the Powerball number 29 is key among the numbers we captured. In fact it is the most prominent number.

Make sure that you show **Lottery Icon** on camera when the press folks come to interview you. I am sure that you can see the possibility of winning more than one jackpot if you follow the methods in **Lottery Icon** . You have studied the book to the territory where you will begin to reap the big rewards.

You need to do one simple exercise before going in.

Go to any past Powerball or lotto game of your choice and print out some past results. Cover any section with blank sheet of paper. Do not look at the result before blocking a section with blank sheet of paper otherwise you will be cheating yourself.

If you see the winning lotto numbers inadvertently flip the printed results down, put the blank paper underneath it and turn it back up.

Follow the process we have discussed so far to capture some winning numbers based on prior results. Use the captured numbers to zero in on the other winning numbers through the Tops Tables. Try and pick total of 18 numbers including the captured ones.

Once this step is done open the printed results and check how many winning numbers from the next lotto drawn result are in your pool. Your goal should be to get five or six winning numbers.

Do this exercise in two or three different places.

If you achieve that, you are ready to go after the multimillion dollar territory. It is more important to get the exercise right. The winnings will follow.

You cannot bet on the lotto compulsively now that you have acquired the knowledge.

I look forward to seeing you on TV with a copy of Lottery Icon of course!

If you want to understand lottery games to the point of becoming a Master, read the next chapter. There is money to be made from that level of mastery otherwise feel free to go out and make some money with the knowledge you have acquired thus far.

If you can afford to put a copy or more of this book in your local library to accord those who may not be able to afford it the opportunity to read the book; I thank you!

Lotto Table

	A	B	C	D	E	F	G	H	I	J	K
R1	0	6	4	7	6	3	9	3	4	9	8
R2	9	2	2	9	0	4	7	8	8	2	8
R3	1	0	1	1	1	9	3	4	0	3	1
R4	5	6	6	2	4	7	7	7	5	6	6

The lotto table is capable of capturing the entire winning lotto trend. It gives you an unbelievable advantage when properly used in conjunction with the rest of the tables.

Let us take a look at some examples with past Powerball results.

You must have laser-beam focus when observing the trends. Capturing just one winning lotto number from the trend could be the one that catapults you to the jackpot.

I am not aware of anybody that has grabbed millions of dollars off the street.

You have to work hard on getting and keeping it as well.

Take a look once more at the winning numbers from the lotto table below.

Lotto Table

	A	B	C	D	E	F	G	H	I	J	K
R1	0	6	4	7	6	3	9	3	4	9	8
R2	9	2	2	9	0	4	7	8	8	2	8
R3	1	0	1	1	1	9	3	4	0	3	1
R4	5	6	6	2	4	7	7	7	5	6	6

In the same time period, Powerball played winning lotto number 56 on 8/15/12. The winning Powerball number 56 is insignificant at this point. There is no way for you to know which trend it belongs to.

You will, however, remember that winning Powerball number 56 is among the groups in the lotto table. It could belong to row R4 columns A and B as well as columns I and J.

The next Powerball result on 8/18/12 played winning Powerball number 1 (one).

In the lotto table, you can see winning Powerball number 10 above 56 in columns A and B.

As I have explained thus far, the winning Powerball number 10 could equally be 01. This would be the time you should use your skill.

Lotto Table

	A	B
R1	0	6
R2	9	2
R3	1	0
R4	5	6

If the winning Powerball number 10 that is above 56 is played as 01, it means that the winning lotto number 92 is likely going to play as 29 to compliment 01.

You will be amazed to know that the next Powerball result on 8/22/12 played winning number 29.

A good student will definitely catch the winning number 29. You would probably think that they might pause at this point.

Oh no, that definitely wasn't the case.

The last winning lotto number among the group in columns A and B is, of course, winning number 6.

The trend continued to yield big money because they played the winning lotto number 6 on 8/25/12.

As you can see, the lotto table trend captured all the winning Powerball numbers.

This amazing feat defies all lotto odds calculation logic.

The trend continued playing rightwards from the very next Powerball result and into the new month.

This time around it started as a decoy. I am exposing you to different ways of understanding and capturing winning trends.

The A and B column trend started with winning Powerball number 56.

The C and D column, on the other hand, started with winning lotto number 47 that actually played as decoy 56. If this sounds confusing, pause for one minute and read the line again.

The winning number 47 is the same as 56. The sum of both numbers is 11 (eleven). If you take one from 7, the second digit, and add it to the first digit, 4, 47 will become 56.

$$47 - 01 = 46$$

$$10 + 46 = 56$$

[Decoy for winning Powerball number 47 = 56]

You should not be overly concerned about it beyond knowing that it helps to identify the trend.

You did not know the initial number 56 trend for that of columns A and B either.

In any event, the trend continued because they played winning Powerball number 11 (eleven) on 9/1/12 followed by winning number 26 on 9/5/12 and of course the fourth Powerball number 29 on 9/8/12 to complete the trend.

You can see once more that the trend captured all the Powerball winning numbers right after each other.

The understanding that 47 could be played as decoy 56 to begin the trend is one of the reasons the unique methods in Lottery Icon cannot be boxed into the confines of lotto odds limitations.

The payoff on that knowledge is instantaneous because you are not expected to know that much.

You can also create endless opportunities that would not exist based on the knowledge you acquired from Lottery Icon.

I have demonstrated the capabilities of the lotto table based on trends that started with the winning Powerball number 56.

Let me dig in deeper and create trends with the same winning number 56 and number 55. I will employ two different dates and still capture winning numbers.

I am in essence proving that you will be unstoppable when you study Lottery Icon.

The first trend will be created with the group in columns D and E of the lotto table.

D	E
7	6
9	0
1	1
2	4

Once you master how to create entry points from the least expected parts of the table, the hidden trends will be exposed, thereby increasing the amount of winnings in your hand.

To create trends with winning Powerball number 56, all I will need to do is convert the first digits into Shadow, thereby turning the groups above into the ones below:

D	E
5	6
6	0
4	1
8	4

You can create similar trends and check them on your own. I will put the above created trend against actual Powerball results and the dates the numbers played.

56 played on 7/21/12.

14 (41) played on 7/25/12.

06 (60) played on 7/28/12.

48 (84) played on 8/1/12.

You can see that the newly created trend captured the winning Powerball numbers right after each other.

Let me do one more from a different section of the lotto table to buttress my point. I will create it this time around with number 55.

I will use columns F and G by changing the two digits into Shadow.

F	G
3	9
4	7
9	3
7	7

The above winning numbers when converted into Shadow will become the new group below,

06

15

60

55

Now let us check the new trend against actual Powerball results.

51 played on 8/11/12 and repeated on 8/15/12 with number 06.

The Powerball winning number 55 played on 8/18/12 and repeated on 8/22/12.

The trend was completed with the second winning Powerball number 06. You will recall that they quietly played the first winning Powerball number 06 on 8/15/12.

The entire group repeated.

The trend gave you that indication of possible repeats with the winning Powerball number 06 from our workout.

You will see these opportunities when you study carefully and exercise patience.

The efficacy of the methods in Lottery Icon is unbelievable.

The methods in this book broke all mathematical codes when it comes to lottery games. It is no easy feat to develop trends that could catch consistent lotto winning numbers spanning over four results. You can have the same opportunity in each and every lottery game. This is unheard of.

Let me touch on one more section from the lotto table before closing the section.

I will close out this section with the winning numbers in columns H and I of the lotto table. I will convert the numbers under column H and leave the ones under column I intact.

Lotto Table

H	I
3	4
8	8
4	0
7	5

The above table after Shadow conversion of the numbers under column H will become the new group below,

04

28

10

55

Let us check the new group against actual Powerball past results.

04 played on 8/1/12

55 played on 8/4/12

28 played on 8/8/12

01 (10) played on 8/11/12

You can see that the trend captured all the four winning Powerball numbers right after each other. It doesn't get better than this.

It is never out of place to look at one winning number for a period of two minutes or so.

You need to extract all the opulence from the winning lotto number in front of you.

If you apply this methodology you should be able to comb out six consistent winning lotto numbers out of a pool of 18.

May good fortune follow you.

There is nobody on this planet that can claim to understand lottery games better than you except maybe those who studied the next chapter.

Chapter 6

Becoming a Master (Practicum)

It took great effort and discipline to study **Lottery Icon** up to this point. The lottery winning numbers require more than mere picking and playing numbers.

You can use any of the methods that I have discussed to work on any lottery game. You can equally use the individual tables to work out consistent winning numbers. You will be able to arrive at the same winning numbers via the trends per the respective tables that will lead you to the winning number you are pursuing.

This of course requires a great deal of time.

You will be exposing the other winning numbers in their respective trends at the same time.

In order to become a master in lottery game you must develop the habit of figuring out why a particular number played. Most of the methods we have applied here relied on using prior two winning numbers to capture the third one.

The master on the other hand will have the skill to capture the winning numbers based on just one prior result. You have to completely think outside the box when it comes to working out consistent winning lottery numbers.

Developing the skill to this level requires a lot more than the information you have in **Lottery Icon**.

You should strive to create walls that the lottery game you are playing could not escape.

Let us develop three quick tables at this point. Each of the last two tables will be from the first one. The table construction will use the same formation that we have been discussing in this book.

I will call the tables Process 11, Process 22 and Process 33.

Process 11

	A	B	C	D	E	F	G	H	I	J
1R	7	5	3	1	0	2	4	6	8	9
2R	7	5	3	1	0	2	4	6	8	9
3R	7	5	3	1	0	2	4	6	8	9
4R	7	5	3	1	0	2	4	6	8	9
5R	7	5	3	1	0	2	4	6	8	9

You will notice that few people, if any, will expect you to form any table in the above format. The numeric order that is taught in schools affect the way people process and write down the information.

The reason I have Process 11 in the above format is to be able to capture the trends regardless of the game.

At the end of Process 22 and Process 33, the numbers would have been repositioned in a way that it will capture all the trends regardless of what is being played.

I will proceed to form Process 22 and Process 33.

I will make more use of Process 33 and use the other two tables if and when necessary.

I do not need to explain the construction at this point.

Let the construction continue.

Process 22

	A	B	C	D	E	F	G	H	I	J
1R	0	4	8	7	0	2	3	7	3	2
2R	2	6	9	1	0	2	9	5	0	5
3R	4	8	3	1	1	8	4	1	7	8
4R	6	5	3	0	6	9	3	6	6	9
5R	7	5	2	4	1	5	7	4	8	9

Process 33

	A	B	C	D	E	F	G	H	I	J
1R	1	4	0	6	0	9	9	7	3	5
2R	8	5	8	7	6	2	9	6	1	8
3R	9	7	8	0	1	6	3	4	2	8
4R	1	4	3	0	6	5	2	7	4	2
5R	0	5	2	3	1	5	4	7	3	9

I have stated several times that all lottery numbers are related. If you understand and take advantage of that knowledge you could work out, and see, numerous winning lottery numbers that others may not be able to see.

You can equally use trends that are not playing currently and zero in on winning numbers. This level of mastery requires more critical thinking, but the reward is enormous.

You should make an effort to include parlay (front and back pair) in your games. There will be many times you will get two out of three numbers on the Pick 3 games as well as others.

The Pick 3 numbers 978 could replace the first or last digit with any other number that could change the rest of the trends.

If you bet 978 and your lottery market plays 971 or 178 you are going to lose the bet unless you include parlay betting as part of your initial plan.

Your 978 could be played as 578 if your market plays the same numbers four games or more earlier, or the same number could play as 378 if they chose to play it four days or so later. Detailed explanation of the two scenarios is beyond the scope of this book.

There are two things I said throughout **Lottery Icon** that I must revisit before I close this chapter.

The first one is that all lottery numbers are related.

The knowledge of that will put you miles ahead of the crowd. If you are interested in being one of the few that are willing to learn more, contact me through any of the big social media. The space will fill up quickly.

I will take only a couple thousand readers. There are times the email could be delayed for whatever reason.

I personally reply all lottery related questions.

The other point that I said several times in this book is that the lottery game uses decoys to throw the lottery bettors off guard. I will touch on that briefly. The more detailed part will be left for the readers who chose to join the few in becoming masters when it comes to lottery.

Let me touch a tiny example through Maryland lottery past results.

Let us use Process 33 above and form a trend with Maryland Pick 3 results of 10/15/12.

Maryland played 424 and 235 on 10/15/12.

I will not go into elaborate explanations since I believe that this is basic for those who studied the book.

Becoming a Master is not for those that are not willing to study.

Maryland played 424 and 235 on 10/15/12.

I will look for the column that will give me 4 and 2 to represent the first digits.

The second or middle digits will be 2 and 3, and the last or third digits will be 4 and 5. Once I get that accomplished it will reproduce 424 and 235.

I will take you a bit further after that.

Please keep reading and pay close attention.

In order to form a trend from Process 33 that has 4 and 2 in the first column I chose column G.

The numbers under column G are the ones below,

G
9
9
3
2
4

I will write down the same numbers in the other direction so that I will have 4 and 2 to represent the first digits of 424 and 235.

The numbers will now be as follows,

4

2

3

9

9

I chose column C for the second digits.

The numbers under column C are as follows,

C
0
8
8
3
2

I will now do the same thing like the earlier diagram from the other direction. The new numbers will now be as follows,

2

3

8

8

0

If you put them beside the ones earlier for the first digits you will get the following set of numbers,

42

23

38

98

90

To get the third digits you will do the same thing with Process 33 column B and the new group including column B will be as follows,

424

235

387

984

905

You have now created trends with 424 and 235 that Maryland played on 10/15/12. One of the things you should consider doing is playing the rest of the group as we discussed so far. You will gain more by including the parlay betting.

You can also extract additional winnings by employing the lottery 1000 Plan.

I have not mentioned this until now.

There is more for those who choose to master more.

The lottery 1000 Plan is when you minus the Pick 3 numbers in the trend from 1000, place them beside the original numbers and apply the three methods, (Numbers, Shadow and Counterpart) to maximize your winnings.

I will place the 1000 Plan Pick 3 numbers side by side with the original trend.

Pick 3	1000 Plan
424	576
235	765
387	613
984	016
905	095

The Pick 3 numbers on the right were formed after the lottery 1000 Plan application. You will not know that 235 is a repeat of 424 unless you use the lottery 1000 Plan.

Becoming a Master requires that you see ahead of the curve.

There are hints that could lead you to some of the winning numbers as a Master.

Go back to Maryland Pick 3 results of October, 2012.

The derived Pick 3 numbers from the lottery 1000 Plan on the right side gave all the clues you need to win some of the Pick 3 numbers on the left side and vice versa. It is a known fact that the Pick 3 numbers on the right are from those on the left after lottery 1000 Plan application.

Maryland played 901 on 10/13/12 followed by 907 on 10/14/12.

The two numbers came from 016 and 095 of the lottery 1000 Plan. If you convert the 6 from 016 into Shadow it will become 019 and if you convert the 5 of 095 into Shadow it will become 097. They made it even easier to see by only converting the third digit. You can see that our unique method captured the proper position regardless. If you have the group as laid out above you will see that the trend is playing upwards.

The next day the trend continued by switching to the original Pick 3 numbers and playing 424 and 235 on 10/15/12.

They completed the trend by playing 873 on 10/19/12.

The reality is that all of the winning numbers were set up before the ones I discussed above.

You may not be able to see beyond this at this point.

You could learn more in the course of becoming a Master.

One bet I can place at this point is that there is nobody on the planet that understands lottery games better than you if you study this book, except for the person who became a Master.

Defy the odds and go win some serious money.

The field is wide open.

Let the game begin.

I cannot over emphasize the importance of adding parlay to your game.

I was in a local Maryland store yesterday (10/21/13) and played some Pick 3 games, including two pertinent lottery tickets to show the readers the importance of playing parlay.

Take a look at two of the tickets I played yesterday to demonstrate the importance of including parlay in your bets.

One of the tickets was played for midday and the other one for evening.

Continue reading after looking at the tickets.

Term: 55354001   21 Oct 2013 09:54
101560045610112-16   142196

 **Pick 3** $0.50 - 1 Draw
                                          10/21
                                         MIDDAY

735   6Box $0.50
                  101560045610112-16

Term: 55354001   21 Oct 2013 19:07
101560270305152-84   395816

 **Pick 3** $0.50 - 1 Draw
                                          10/21
                                         EVENING

333   Straight $0.50
                  101560270305152-84

In the first ticket Maryland played 8 in place of 7. Maryland midday for 10/21/13 was 538 against the 735 of the above ticket.

I did something similar to show the readers for the evening result. The Pick 3 number for the second ticket is 333 while Maryland played 332.

In both cases you will lose unless you include parlay in the bets.

You would rather make $100 or more than end up losing the bet. You may think that the amount is too small but believe me it is not. If you multiply only that amount daily over one year you will see that it could make good change in a lot of pockets.

Beyond Random Numbers

The common axiom on random numbers does not mean that one cannot do close examinations on them.

The question has not been properly asked or answered up to this point.

The properly phrased question is,

Is there such a thing as a random number?

The answer should not be vague.

To avoid answering it lightly we must put it to test.

I will create some super random numbers and put them against actual lottery results.

I use the word super because the numbers here are manually written down by me. They are better than the computer generated ones because I have complete control of the decision. I encourage you to do the same exercise with the same super random numbers or any other sets created by you after studying this book thoroughly.

I will form the super random numbers from 0, 1, 2, 3, 4, 5, 6, 7, 8 and 9.

The first sets I decided to form will be 5 numbers in columns and 5 numbers in rows. I am starting from any point and decided to place number 8 before the others. I chose to put down number 8 three times because I have complete control of the placement. I will continue with the rest of the numbers.

I formed the second set below that.

The super random numbers are as follows,

88820

37924

86641

09310

29750

The second set of super random numbers is as follows,

0283

1374

4565

2292

0124

I will now use the above numbers to construct super random tables 1 and 2 before putting them to the test. I used the same format that we have been going through over the book. You should be able to do this at this point, otherwise go through the book once more.

Super Random Table 1

	A	B	C	D	E	F	G	H	I	J
R1	0	2	1	3	0	6	6	2	7	8
R2	0	3	1	2	5	3	9	6	4	9
R3	9	3	8	4	4	2	7	2	8	2
R4	8	8	4	1	2	0	1	9	8	4
R5	8	7	0	5	5	0	2	8	2	0

Super Random Table 2

	A	B	C	D	E	F	G	H	I	J
R1	4	7	4	8	0	0	1	8	8	0
R2	2	6	2	3	2	6	4	3	5	3
R3	9	8	1	1	2	8	6	5	0	2
R4	2	2	4	5	9	9	0	2	8	8
R5	0	8	3	1	4	7	9	2	7	0

I will now use sections of the Super Random Tables to test the veracity of random numbers as pertaining to lottery.

Random Numbers exist as long as you are within the luxury confines of the numeric numbers taught in schools. If you make adequate use of Numbers, Shadow and Counterpart that I have been teaching through this book, you will find out that the word random is meaningless. You can in fact use any set of numbers and zero in on the winning numbers.

It works outside the walls of the numeric numbers.

Let us examine one example. I am sure you know by now that I often use my entry point to identify trends from the least expected angles. The trends are often hidden in those places but could easily be exposed when you learn to apply the methods.

Let us look at the Super Random Table 2 against actual Maryland lottery results.

Maryland Pick 3 played 812 on 10/16/12.

This method is so powerful that it actually exposes the trends with just one result.

I decided to create the winning Pick 3 number 812 from columns F, G and H of Super Random Tables 2.

Let us bring that table down before the conversion that will create 812.

F	G	H
0	1	8
6	4	3
8	6	5
9	0	2
7	9	2

Let me write down the same group of Pick 3 numbers without the grids and proceed to convert them afterwards.

018

643

865

902

792

In order to have number 8 in the first digit of 018 instead of 0 (zero) I will have to convert the 0 (zero) into Shadow (3) and then into Counterpart (8). I will do the same thing for all the numbers under the first column i.e. numbers below the 0 (zero).

I will leave the second (middle) numbers intact and proceed to change the third (last) digits into Shadow so that the 8 of 018 will become 2 thereby creating 812. You must also do the same conversion to all the Pick 3 numbers in the group so that the trend will remain intact.

The finished group will go from the ones above to the newly formed ones below,

812

440

767

108

098

I am confidently doing this because all numbers are the same contrary to popular belief. I have stated this numerous times in my books.

Let us examine the newly created trend against Maryland lottery results of October, 2012.

Maryland played 812 on 10/16/12.

Maryland played 440 on 10/21/12.

Maryland played 098 on 10/22/12.

You can see that the trend created by 812 is playing in Maryland. Three out of a total of five Pick 3 numbers played so far. Those winning Pick 3 numbers could have made serious money for you because they played straight.

The two winning numbers in the trend that are yet to play are 767 and 108. Now spring forward to the next month. You will notice that Maryland Pick 3 played 767 on 11/14/12 and 801 on 11/15/12. This is simply incredible.

Take another look at the dates the Pick 3 numbers in the trend played.

The dates that the numbers played are 11/14/12, 11/15/12/, 10/16/12, 10/21/12 and 10/22/12. I could go into the significance of the dates and zeroing in on the winning numbers with the dates. That is beyond the scope of this book and will indeed make it cumbersome.

The most important fact here is that I worked out actual winning trends with the so called random numbers.

You must pay close attention to see when the trend is being altered and follow suit. Any slight variation actually creates additional winnings for good number scholars. The reason for that is because you are not expected to see those numbers.

Let me take the next Pick 3 number that played on the same date with 812 of 10/16/12 to demonstrate this. That winning Pick 3 number is 755.

I will create the conversion with Pick 3 numbers from Super Random Table 2, columns D, E and F.

D	E	F
8	0	0
3	2	6
1	2	8
5	9	9
1	4	7

I will convert the 8 of 800 into Shadow (2) and then into Counterpart 7. Do not forget to do the same conversion to the rest of the numbers under column D. I will proceed to convert the 0 and 0 of columns E and F into Counterpart 5 and 5 to create the Pick 3 number 755. I will do the same thing to the rest of the group of which the newly created group will appear as the ones below,

755

571

973

244

992

Take a very close look at the new trend and dates the group played. I want you to carefully observe when they made minor changes. The objective of this is for you to notice when such happens and capture the winning numbers.

Maryland played 424 on 10/15/12.

Maryland played 755 on 10/16/12.

Maryland played 929 on 10/20/12.

873 (973) played on 10/19/12

527 (571) played on 10/20/12

You can see that they minus one from 973 to make it 873 and added it to the one (1) of 571 to make it 527. A good observer could capture the 527 and actually win it straight. You should learn to patiently look at the individual Pick 3 numbers and what is going on.

You may be tempted to wonder why I should make the bold statement that you could indeed win the 527 straight.

You will notice that the number 2 played in the middle for the winning numbers 424, 929 and of course 527 will not be left out. The 424 and 929 are good indicators that the number 2 is likely going to play in the middle.

The 7 on the other hand played in the first digit position for 755 followed by the 7 on second (middle) digit position for 873 and will complete that trend with the 7 for third (last) digit position on the 527. The position of the two numbers 7 and 2 are given thereby making it easier for you to properly line up the winning Pick 3 number 527.

755

873

527

The last winning Pick 3 number in the trend 527 captured the three lottery number movements, Numbers, Shadow and Counterpart.

This once more is done through the so called random numbers.

You can begin to see the answers to the question about the veracity of random numbers.

The most lucrative thing you can do with Super Random Table is to create consistent winning parlay numbers. As you get better you will begin to make necessary adjustments and capture Pick 3 and Pick 4 numbers as well.

You can decimate every lotto game odds.

Let me show you one opportunity that presented itself recently through Maryland lottery Pick 4 games.

You will master this method over time. I will make more information available at the end of this book.

I have stated several times in this book that all numbers are the same. It might not sound proper because nobody taught you that in school.

Maryland lottery played 4073 on 11/31/2013.

There are several ways that I can zero in on winning lottery numbers regardless of what is being played. Let me show you the trend that led to that winning lottery number without big elaboration.

Maryland recent Pick 4 winning numbers,

0585 and 0810 played on 10/30/13

7008 and 4073 played on 10/31/13

Now take a look at the trend that produced the winning Pick 4 number straight.

4452

4073

3027

0421

1575

You may be pondering on what the correlation could be between the Maryland Pick 4 numbers above and my trend beyond the known winning Pick 4 number 4073. The reason for that is because you are not trained to see it. The winning numbers are hidden.

How did I get to the winning Pick 4 number 4073?

Maryland played 0585 while my trend produced 1575. The two Pick 4 numbers are the same. If you take the number one (1) from my trend and add it to the number seven (7) it will produce 0585 that Maryland played.

Maryland came back the next day and played 7008.

My trend calculation produced 4452 against 7008.

You can see that if you add the first (4) and second (4) digits you will get 8. If you add the third (5) and the last (2) digits together you will get 7 thereby producing the Pick 4 number 7008.

My trend clearly produced 0585 and 7008.

The only thing that is left would be to bet on the next winning Pick 4 number in the trend. That Pick 4 number is, of course, 4073.

That was a wonderful day. The month ended Superb.

The next winning Maryland number is 4073 straight.

You can create the same opportunity in each and every single game.

Let the skeptics spend their time talking about the impossibility of not winning the lottery while you spend yours catching the winning lottery numbers and smiling to the bank.

If you are ready to become a master beyond this book, drop me a line via most of the social media.

I will make more information available at the end of this book

There are times the emails get delayed for reasons beyond me.

I do reply to all lottery related questions personally.

You gain a whole lot more by studying **Lottery Icon** several times. It is a different language that is worth the effort.

Nothing good comes without effort in life.

Believe me, you should be able to master this book and indeed become a genius.

This will give you further proof that you could enjoy endless winnings based on the knowledge you acquired from Lottery Icon.

I do vehemently believe that all numbers are the same.

I will buttress than assertion further by taking you through angles that you might not have imagined and show you that you can actually create more winnings.

I also mentioned several times that parlay betting should be part of your betting games.

I will take different games from different dates including Pick 3, Pick 4, Pick 5, Multi Match, Mega Millions and numeric numbers and prove that you can add more winnings in your pocket.

Once you properly align the numbers using my methods you will catch the trends and bet for the big money.

The parlay pays about $50 on $1 winnings.

Six way Pick 3 lottery numbers like 123 will cost you about $12 to bet on the parlay.

Three way Pick 3 lottery numbers like 223 will cost you about $6 and pays the same $50.

In the case of six way lottery numbers you would have to bet on more than four different parlay numbers to lose against any potential winning of $50.

You would of course start at a disadvantage without the knowledge I am about to expose you to.

You will on the other hand be at a huge advantage with this knowledge.

The reason for that is that you will not engage on the betting with this method until you properly align the numbers from the table against the actual trend you will be betting on.

Once you properly align the table with the trend that your local market is playing it will produce consistent winnings.

You can use this method through years of betting.

You can align different trends that will also produce more winnings.

Further explanation without actual proof at this point will make this argument cumbersome.

There is no lottery game you cannot win with application of my unique tables, Numbers, Shadow and Counterpart.

Let us begin the Process by picking different games from different states that will encompass different dates.

The first pick will be from Maryland lottery.

Pick 3

050 and 413 played on 11/16/12.

Pick 4

0277 and 4662 played on 11/14/12

Maryland Multi Match

| 07 | 16 | 18 | 23 | 35 | 40 played on 11/15/12 |

Mega Millions

| 11 | 28 | 33 | 41 | 43 | 41 played on 12/14/12 |

DC 5

98624 and 46404 played on 1/5/13

Numeric Numbers

0 and 1 added to make it a total of 40 numbers.

I will be creating tables that will take a total of 50 numbers. I decided to stop at 40.

You can clearly see that I chose different games from different states with different dates.

I will build tables with the process we have discussed throughout **Lottery Icon**. I will bring down the above numbers without the name of the games and proceed to build tables.

I will build one original table from the above and repeat the process two more times. I will call the tables P1, P2 and P3.

You may choose to use the tables or create yours like the ones I am going to build.

The major thing I want to do here is show you how to realign the table trends and capture consistent lottery winning numbers in your local market.

The Numbers

050 413

0277 4662

07	16	18	23	35
11	28	33	41	43

98624 46404

0 1

I will now begin to use the above numbers to construct the tables.

You are now beginning to get into the advanced stage where I do not spend too much time on explaining the process of constructing the tables

P1

A	B	C	D	E	F	G	H	I	J
4	1	1	3	1	3	6	6	3	4
0	6	8	4	4	3	8	6	6	3
1	2	0	1	2	9	0	4	4	0
1	5	4	0	1	8	4	2	4	7
0	3	7	4	5	2	2	0	7	1

P2

A	B	C	D	E	F	G	H	I	J
2	0	6	1	1	8	8	0	0	2
9	6	0	3	1	3	7	6	5	2
8	4	1	0	4	4	6	4	0	7
4	1	4	4	2	3	7	6	0	4
4	5	3	4	1	3	1	2	3	1

P3

A	B	C	D	E	F	G	H	I	J
4	6	5	4	1	3	0	4	1	3
4	6	7	1	2	8	4	6	1	4
7	0	6	4	3	0	8	4	9	3
4	0	4	1	6	2	3	0	2	2
2	1	3	7	0	5	8	1	0	1

Align to your advantage before betting on parlay or any other game of your choice. This method works excellent on parlay as well as other lotto games.

The winning lottery numbers are made for you not to figure out easily. You will begin to learn how to properly align the winning numbers in this introductory advanced stage.

Let us work on one through New York Numbers.

I will begin this process with 637 and 502 that played on 10/1/13 in NY.

The basic alignment will be to form the two winning Numbers that began the month of October 2013. I will begin by looking for lead numbers that will produce 6 for 637 and 5 below that for 502. I will do the same thing for both Pick 3 numbers.

My best entry point will be from P3, column E.

I chose that because I do not have 6 and 5 readily bearing in mind that I could choose any entry point since I have the advantage of Numbers, Shadow and Counterpart.

The numbers under P3, column E are,

1, 2, 3, 6 and 0.

I can achieve my goal of 6 and 5 by applying Counterpart which would change the above numbers into the new ones below,

6, 7, 8, 1 and 5 which could be read in the desired trend as,

6, 5, 1, 8 and 7.

You can see that I have achieved my initial goal of having 6 and 5 close to each other for the first digits of 637 and 502.

The middle digits will require 3 for 637 and 0 for 502. I have to find or create it from the tables. This search for the middle digit numbers led me to P2, column D that has the following numbers,

1, 3, 0, 4 and 4.

You can see that I already have 3 and 0 close together. I will move the number 1 (one) to the last position so that I can have 3 and 0 (zero) to align with the initial formed numbers (6, 5, 1, 8 and 7) for the first digit positions.

The second digits above after moving number 1 to last position will now appear as follows,

3, 0, 4, 4 and 1.

I have formed numbers that will cover the first and middle digits for 637 and 502.

6, 5, 1, 8 and 7 for the first digits.

3, 0, 4, 4 and 1 for the middle digits.

The third digits must have 7 and 2 to complete the process of creating 637 and 502.

The search for the third digits led me to P2, column J that has the following numbers,

2, 2, 7, 4 and 1 which I will read from the number 7 in the other direction that will give me the required 7 and 2 for the third digits.

The numbers will now appear as follows,

7, 2, 2, 1 and 4.

I have now completed the process of creating numbers that will give me 637 and 502 that played in New York on 10/1/13.

I have formed numbers that will cover the first, middle and third digits for 637 and 502.

6, 5, 1, 8 and 7 for the first digits.

3, 0, 4, 4 and 1 for the middle digits.

7, 2, 2, 1 and 4 for the third digits.

If you read the newly formed numbers from left to right you will see 637 followed by 502 and the rest of the trend. Let me position the new groups in the lottery result format that would make it easier for some scholars. Only scholars read to this point.

The newly formed trend is,

637

502

142

841

714

Now let us see what happened with New York Numbers (remember that your are aligning the numbers to enable you capture more winning numbers.

New York Numbers played 637 and 502 on 10/1/13 followed by 142 on 10/4/13. The next winning Pick 3 number 142 did not play until four days later.

The following Pick 3 played,

014 and 249 on 10/2/13, 058 and 188 on 10/3/13 followed by 331 and 142 on 10/4/13.

The real question is,

Why did other numbers play before 142?

Is there significance to the other numbers that played prior to 142?

Do the other numbers give hint to what is coming?

Could you take advantage of that knowledge and capture more winnings?

The answer to the above questions is YES.

A lot of lottery players will play 142 right after 637 and 502 and give up. They may not see the need to play 142 for the next three days.

Let us examine the significance of the other numbers that played in between the trend we created.

The trend we have so far is,

637

502

142

841

714

The next winning Pick 3 number that played after 637 and 502 is 014.

You will look at the above trend that you have already created and see if you could create the next number that just played 014. You would have won some money with the 142 anyway by playing the parlay. You will not be satisfied as a scholar which is why you started entertaining the idea of recreating 014.

You can achieve that goal by converting the first digits into Counterpart and then Shadow to enable you change the 8 of 841 into 0 (zero) thereby making it 041. You will do the same to the rest of the groups to maintain the trend.

The numbers in the first digits are,

6, 5, 1, 8 and 7 that will be changed to Counterpart,

1, 0, 6, 3 and 2 that will be changed to Shadow,

4, 3, 9, 0 and 8 as the new numbers for the first digits.

I will put down the original trend and place the new group with 014 beside them.

637	437
502	302
142	942
841	041
714	814

You can see that the newly formed 041 exposed 942 in the same position as 142. The average player will play heavy on 142 while you exposed the 942 opportunity that played together with 041 on 10/2/13.

You can see how I am beginning to align the winning numbers and exposing more opportunities at the same time.

The next day New York Numbers played 058 and 188 on 10/3/13.

You job should be to expose one and capture the next winning Pick 3 numbers. The naysayers will not see this opportunity.

I will do the job for this time.

You will take another hard look at the trends you have just created and see where if possible you can create the potential opportunity.

The new Pick 3 number in your hand is 058.

You may not know how to capture 058 at this point but you definitely can use it and capture the next winning number 188 that played with 058.

The goal is to create 058 with the working trend in your hand.

Let me bring down the original trend before embarking on creating the trend with 058.

637

502

142

841

714

You can create the 0 and 5 of 058 by just changing the first two digits into Counterpart. If you start from 502 and 637 you will create,

05

18

26

39

69

The above numbers would accomplish the goal of creating 0 and 5. You will now complete the third digits by converting the group into Shadow. The reason for that is to turn the number 2 into 8. You should consider making use of each of the 2's or playing parlay. Remember that they could go either way. If you do this process with either of the two's you will still create winning trends. The reason for that is because all numbers are the same with the application of Numbers, Shadow and Counterpart.

Let me just use one direction for convenience at this point.

The third digit numbers are 7, 2, 2, 1 and 4 that I will now convert with Shadow application into,

5, 8, 8, 4 and 1 that I would in turn move the first number 5 into the last position thereby having the group as,

8, 8, 4, 1 and 5.

You must remember that I did this earlier and I am now reversing it to capture the 058 trend.

The two digits that we have already formed are,

05

18

26

39

69

I will now add the third digits to create the Pick 3 trend based on the new winning Pick 3 number 058.

The group including the newly formed last digits will be as follows,

058

188

264

391

695

You will notice that you have now captured the next winning New York Number 188 that played with 058 on 10/3/13.

For the bettors who doubt the efficacy of the method, 264 played right after 139 on 10/6/13 and 10/7/13/.

You can see how I use the original trend to consistently follow their trend and capture winning lottery numbers. You can do this for midday and evening. You can create more than one way of capturing the winning lottery numbers.

By the way I did not forget the other winning Pick 3 number 133 that played on 10/4/13 with Pick 3 number 142.

Your assignment is to figure it out. If you do it right the other winning trends will be exposed. If you do it wrong go back and study **Lottery Icon** further.

The payoff is worth the effort.

It is definitely amazing that I can do this with different lottery games.

Private Club Betting

I studied numbers for so many years to a depth that many could not imagine. This journey took me to several continents. I also developed methods that capture winning numbers like no other.

I also know that if you truly study Lottery Icon you will be ahead of the crowd.

I wish that I could include all of the methods in Lottery Icon.

I do, however, understand that a lot of readers will not stand the idea of a book being voluminous, especially one that is dense in content.

I will furnish the rest of the materials to the readers that are willing to go beyond this book.

I anticipate working with a group of two thousand ONLY. The class will fill up quick.

That group will get more than the information in this book including mastering how to enjoy consistent winnings with Private Club Betting.

Private Club Betting is one of the ultimate systems that is beyond the scope of **Lottery Icon** and will be taught in the advanced stage.

This ultimate system catches every lotto game. You will be thinking outside the box once you understand it.

They will also become masters when it comes to betting on lotto games including the Mega Millions where the odds adjusted upwards recently to the tune of 259 million.

There is more to learn and Private Club Betting is just one of them.

 I will give you a snippet of Private Club Betting in action.

Pick 3

711 Maryland played 711 on 12/22/13

007 Maryland played 007 on 12/23/13

080 learn how to win 080

063 Maryland played 306 on 12/24/13

740 Maryland played 470 on 12/22/13

Pick 4

3692 Maryland played 3692 on 11/7/13

5640 Maryland played 4605 on 12/13/13

8634 Maryland played 3648 on 12/11/13

3821 Maryland played 3821 on 12/18/13

0240 Maryland played 0204 on 12/23/13

You spend more to make more.

If you think for one minute that the lottery could not be won every single day, this class is for you.

You will also have all the information that is not in Lottery Icon.

The class will not take anybody beyond the first two thousand. I expect it to fill up fast.

Contact me through the social media for more information.

XXXXXXXXXXXXXXXXXXXXXXXXXXXXXXXX

MS B

Good to read from you.

There is no difference between the Pick 3, Pick 4 and lotto games contrary to popular belief.

Actual Examples

Pick 3	Pick 4	Pick 5	Lotto
012	0123	01234	01, 2, 3, 4, 5, 6

Lotto Expanded Field

*** The same lotto numbers above could be expanded to*** 12, 23, 34, 45, 56***

The above winning lotto numbers could also skip some numbers and completely move away from the numeric order.

The above lotto numbers could produce the following numbers,

12, 13, 14, 15, 16 with 1(one) being the lead number.

It could also do the same thing with number 1(one) being in the second (last) digit position.

This same trend applies to all of the numbers beyond the ones I am using as example so far.

All the numbers will form trends in their respective positions.

*** The same lotto numbers above could be expanded to*** 12, 23, 34, 45, 56***

If I create trends from the above winning numbers and replace the second digits I will have the following numbers,

11, 21, 31, 41, 51.

These numbers are also forming trends.

You could identify the winning trends when you apply the methods from Lottery Little Book.

I will now touch on the Powerball results with the last book you read.

I will use the last three Powerball results that ended the month of September and go into October where that would give you complete one month Powerball results.

The reason is to show you that there are indeed relationships between prior and current results.

Powerball Results

12	17	45	54	58	13 (9/21/13)
2	7	17	49	53	23 (9/25/13)
14	47	52	53	54	5 (9/28/13)

New Month of October First Powerball Result

| 4 | 6 | 25 | 42 | 51 | 17 (10/2/13) |
| 11 | 12 | 17 | 39 | 40 | 5 (10/5/13) |

You will notice that numbers 12 and 13 played on 9/21/13. The next winning number based on that trend would be either 11 or 14. The next result produced winning number 2 instead. It means that you should expect number 1 or 3 to also play. They would follow the precedent set by 12 and 13. The number 12 and 13 by the way is among the numbers I created in the earlier part of this article.

The next result produced winning number 14 to compliment the 12 and 13 that already played. You would naturally expect the next winning Powerball number to be 15. They switched the mode by playing the trend that placed 1 (one) in the last position thereby playing the number 51 instead of 15.

You have more than one way of identifying the winning number 51.

In the above results, you can clearly see the trend with the winning numbers 54 on 9/21/13 followed by 53 on 9/25/13 and 52 on 9/28/13.

The next winning number based on that trend would of course be 51 to begin the month of October.

You could easily have identified winning number 51 as one of your play numbers.

In the above results you will notice that the number 17 has major presence. It did not happen by accident. You will also notice the number 45.

The Shadow of 45 is 17. The Shadow of 7 is 5 and the Shadow of 47 is 15 that played as 51. You will notice that the Shadows played in the next result which would buttress the position of choosing 51 as potential play. It shows you that the related numbers are going to be playing through the new month. If you do carefully study of the basic numbers I started with, you will notice how they fit in the Powerball actual trend.

Now let me put down the related numbers with just the winning number 51. You should bear in mind that you can use the same application for all the numbers to enable you have idea of what is coming next.

The numbers 51 conversions will be in the order of Number, Shadow and Counterpart.

51, 74, 29, 86, 31, <u>04</u>, 59, 76, 21, 84, 39, **06**

The underlined numbers played on 10/2/13 to begin the month. The numbers are related as you can see through Shadow and Counterpart.

The next winning Powerball number played the next number 39. You will find 39 beside the winning number 06.

You would expect the next winning Powerball numbers to produce 21 and 59 based on the trend.

They actually played 21 and 59 to start the trend. The numbers are not meant for you to be able to identify them easily.

They played them and skipped one month. The average lotto bettor would not see this trend. There is a reason I wanted you to read Lottery Little Book.

The winning Powerball number 21 played on 8/3/13 followed by 59 on 8/7/13.

A good lotto number scholar would tell you that the above two set the trend that I am discussing here.

The winning Powerball number 31 played on 10/23/13 followed by 29 on 10/26/13.

You can see how the number 51 produced the winning trends in Precision Order.

Study the books and make solid use of them.

Be prepared to read the final edition when it comes out.

You will see how I decimated every lotto odd calculations in the upcoming book.

Keep on practicing on paper until you become best.

You should try to get six winning numbers and add additional 12 thereby making it a total of 18 solid numbers using my methods.

Wheel the 18 numbers with the free lotto wheeling systems online that would give you solid total 42 plays that would help you catch the GOLD.

Oh by the way; I am glad to read that you captured all five numbers on paper in the California Fantasy 5 with the methods outlined in the number one lottery book,

Lottery Little Book.

Final Note

I thank all of my supporters.

Lottery Icon has unveiled the wall of lottery .

I sincerely urge you to read **Lottery Icon** more than once or at best study it thoroughly.

The advanced course will be available in the form of a book (paperback).

The book will cost about $100.00 (non-refundable).

For more details, contact me through social media.

Lottery Icon could not be possible without the Power of The Superintendent of the Universe,

The Almighty God.

The Most Precious and Priceless Good that is and will always remain available to Good People is the TRUTH.

Always seek the TRUTH relentlessly, Embrace the TRUTH warmly, Live the TRUTH genuinely and Diseminate the TRUTH lavishly.

Eze Ugbor

www.ingramcontent.com/pod-product-compliance
Lightning Source LLC
Chambersburg PA
CBHW080242180526
45167CB00006B/2387